# FLORA OF
# THE OUTER HEBRIDES

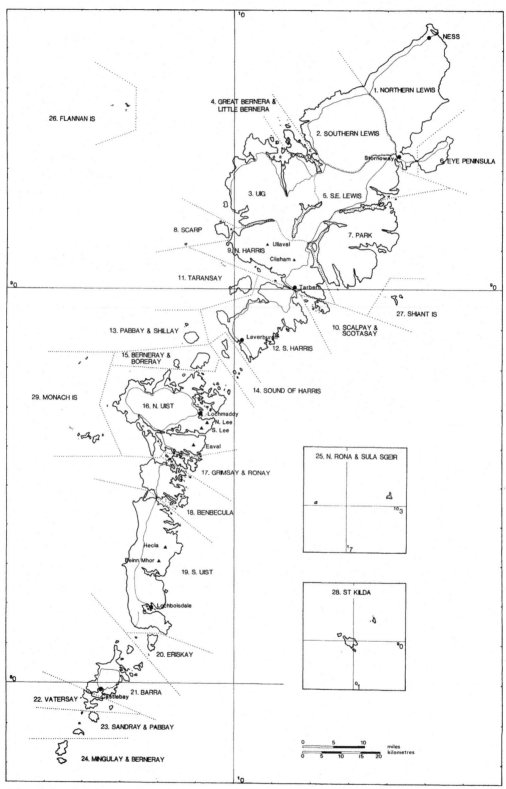

NESS

1. NORTHERN LEWIS

4. GREAT BERNERA &
LITTLE BERNERA

2. SOUTHERN LEWIS

26. FLANNAN IS

Stornoway

6. EYE PENINSULA

3. UIG

5. S.E. LEWIS

8. SCARP

▲ Ullaval

7. PARK

9. N. HARRIS

Clisham ▲

11. TARANSAY

Tarbert

27. SHIANT IS

13. PABBAY & SHILLAY

Leverburgh

10. SCALPAY &
SCOTASAY

15. BERNERAY &
BORERAY

12. S. HARRIS

29. MONACH IS

16. N. UIST

14. SOUND OF HARRIS

Lochmaddy
▲ N. Lee
S. Lee

▲ Eaval

17. GRIMSAY & RONAY

25. N. RONA & SULA SGEIR

18. BENBECULA

Hecla ▲

Beinn/Mhor ▲

19. S. UIST

28. ST KILDA

Lochboisdale

20. ERISKAY

22. VATERSAY

Castlebay

21. BARRA

23. SANDRAY & PABBAY

24. MINGULAY & BERNERAY

miles
kilometres
0   5   10
0   5   10   15   20

*Map of the Outer Hebrides*

# FLORA OF THE OUTER HEBRIDES

**R.J. Pankhurst and J.M. Mullin**

The Natural History Museum
www.nhm.ac.uk/publishing

Pelagic Publishing
www.pelagicpublishing.com

First published 1991 by The Natural History Museum
© The Natural History Museum, London, 1991

This edition printed and published by Pelagic Publishing, 2013,
in association with The Natural History Museum, London
www.pelagicpublishing.com
www.nhm.ac.uk/publishing

Pelagic Publishing, PO Box 725, Exeter, EX1 9QU, UK

ISBN 978-1-907807-49-7

This book is a reprint edition of 1-898298-84-X
Original cover design by Michael Morey

Front cover illustration: *Dactylorhiza incarnata* ssp. *coccinea* growing on wet
machair on North Uist (photograph by Sidney J. Clarke, ARPS, 14 June 1990).

# Contents

# Preface

The Botany Department of the Natural History Museum, London, having completed their impressive investigation of the flora and vegetation of Mull, have since 1979 been engaged in a study of the flowering plants of vice-county 110 with the eventual aim of preparing a modern Flora of the Outer Hebrides. A great deal of recording and collecting in that area has been undertaken by a succession of workers over the past century or more and especially over the last five decades. Professor J. W. Heslop Harrison, for example, wrote in 1939 that for five successive seasons members of his Department of Botany at King's College, Newcastle upon Tyne, had been working in the Hebrides with the intention of producing a County Flora of v.c's 103, 104 and 110. *A preliminary Flora of the Outer Hebrides*, edited by Professor Heslop Harrison, did in fact appear in 1941, but it was not followed by a more definitive publication.

In 1936 Miss 'Maybud' Campbell visited several islands of the Outer Hebrides, making extensive notes on her botanical observations and collecting a large number of plant specimens. She was already a 'semi-permanent voluntary worker' under A. J. Wilmott in the Herbarium of the Natural History Museum. The Hebridean records and plant specimens were therefore deposited in the Museum and were examined and identified by Wilmott. In subsequent years he accompanied Miss Campbell on several further visits to the Outer Isles and assisted in preparing accounts for publication. *The Flora of Uig* (Buncle, Arbroath; 1945), edited by Miss Campbell and with extensive ecological and taxonomic notes by Wilmott, was the most ambitious of these publications. Miss Campbell continued to make frequent visits to the Outer Isles up to 1980, often accompanied by others with scientific interest in the Hebridean flora. She died in 1982 at the age of 79, but had, unfortunately, still not revised *The Flora of Uig* nor written a projected *Flora of Harris*, and her early plan for a Flora of the whole of the Outer Hebrides remained an unfulfilled dream. All her records and thousands of plant specimens have, however, been made available to the Museum's Department of Botany and they undoubtedly constitute a very impressive body of material for use in the preparation of a comprehensive Outer Hebridean flora. It should be added that Andrew Currie of the Nature Conservancy Council, who has now succeeded Miss Campbell as Vice-County Recorder for v.c. 110, has recently (1979) published a *Provisional Checklist of the Vascular Flora of the Outer Hebrides*; 'provisional' because 'it does not include unpublished information in the hands of Miss Campbell', but nevertheless extremely valuable.

This, then, was the unsatisfactory position when the Museum embarked on the task of examining and assessing the great mass of material relevant to a long-needed Flora of the Outer Hebrides, much of it now in the Museum's keeping, and of reducing it to usefully published form. The present publication is the first outcome of this further study of some of the more distant and less accessible parts of the British Isles. In producing it the staff of the British flowering plant section at the Museum have worked together under the guidance of R. J. Pankhurst and have again demonstrated the great value of this kind of co-ordinated effort. The result will certainly be much welcomed by all interested in details of the geographical distribution of British flowering plants, in the extent and nature of the variation they exhibit and in likely explanations of these facts.

A. R. Clapham
1st March 1986

# Introduction

by R. J. Pankhurst

This Flora is intended to be a concise and informative summary of what is so far known about the vascular plants of the Outer Hebrides. Both published and unpublished data has gone into its preparation. Much data already existed in herbarium collections and in numerous scattered and un-coordinated publications. We wish this publication to be both a tool for immediate use and also a basis for a future more detailed and comprehensive Flora.

In his Preface, Prof. Clapham has already explained some of the historical circumstances which make it particularly appropriate for the British Herbarium of the Natural History Museum, London to produce a Flora of the Outer Hebrides. A. J. Wilmott (former head of the British Herbarium at the Natural History Museum), Miss M. S. Campbell (for many years BSBI Recorder for the Outer Hebrides) and their collaborators had built up a collection of about 15,000 sheets of vascular plants in the herbarium as well as a virtually complete collection of all the relevant publications. After Miss Campbell's death, her notebooks for her own Flora project came to the Museum as well. We do not have any explanation of why neither the British Museum team nor Prof. J. W. Heslop Harrison's group at Newcastle never completed their Flora projects, but clearly the Museum was in a good position to be able to round off the work already done. The Outer Hebrides, Watsonian vice-county number 110, is one of the relatively few vice-counties in Britain to have never had a county Flora. The *Flora of Uig* (Campbell, 1945) covers only part of Lewis, and the floristic section is simply an annotated checklist. A. J. Wilmott's contributions on the vegetation and systematics is actually much more substantial. Prof. Heslop Harrison's preliminary Flora (Harrison, 1941a) is comprehensive, but is also an annotated checklist. Its usefulness is reduced by the fact that a few of the records are known to be inaccurate, which unfortunately casts doubt on the many probably genuine but unconfirmed records there published.

In order to give ourselves first-hand experience of the flora of the islands, to collect records from under-recorded areas, and to study the critical groups, three field excursions were organised, in 1979, 1980 and 1983. Special attention was paid to collecting *Euphrasia, Rubus, Hieracium* and *Taraxacum* for determination by experts. Examination of the previous records showed that the outer islands such as St. Kilda, the Monach Islands, or the islands south of Barra, had been rather thoroughly explored (see bibliographies in Appendices), whereas large areas of the mainland had been scarcely touched. This remains substantially true even now. The centre and east coast of north Lewis is still mainly unknown (although there is good reason to think that there are not very many species to be found there). Much of the hilly area between Tarbert and Uig in central Harris is under-explored, and so is the east coast of South Uist, since both these areas are rather inaccessible. The complex coastline, and the innumerable islands large and small, make it impractical to record the Outer Hebrides on a grid square system, given the number of records actually available. We have therefore devised a scheme of 29 numbered zones (see the map on the end papers, and the list in the appendix). These are an attempt at a compromise between the botanical diversity, the geography and the sizes of the areas. It was decided not to attempt to produce dot maps of species distribution because of the geographical difficulties and because there are not yet sufficient records to justify it.

The text of the Flora has been produced on microcomputers by word-processing, and lists of records of species from various areas have been inverted to produce lists of species with the numbers of the areas by means of a computer program. The keys to critical groups were prepared using the interactive key construction program written by RJP. This program is part of the PANKEY package of computer programs for identification. However, the records themselves are entered on a conventional card index and not in a database. This was partly because there was no suitable equipment available when our project started. Also, a record database is not so necessary when the computer is not going to be used to make dot maps.

A card index gazetteer of all place names on the Ordnance Survey 1:50000 map series was prepared by Mr. J. Rogerson and is stored in the British Herbarium. This has proved extremely useful; tracing the name of any one of the innumerable lochs would be impossible without it. The gazetteer was used to help prepare an index of all the place names which appear in the text (see the Appendix). A number of other important herbaria were searched for records as well as our own. A list of these is given in the Appendix. Unfortunately, many of the specimens presumably collected by Prof. J. W. Heslop Harrison have never been located. Some of his duplicates are at Kew and at Cambridge, but the whereabouts of the bulk of his collection remains a mystery. Part of the collection of his colleague W. A. Clark, who took part in many of the Newcastle expeditions to the Outer Hebrides, was donated to the Museum, but unfortunately, this proved to be mostly too decayed to be of any value. For convenience, a current list of National Nature Reserves and Sites of Special Scientific Interest is also provided, as well as list of all the collectors that we have mentioned in the text.

The project has been managed and organised throughout by RJP, who also planned and took part in each of the field expeditions. The first draft of the species account was prepared by JMM, and this was then edited and extended by RJP. All the keys to critical groups were written by RJP. Ursula Preston and Mary Chorley did most of the work of filling in the index cards from our herbarium and from literature. Andrew Currie took over the task of vice-county recorder from Miss Campbell and has assisted us with the project throughout. Mrs. C. W. Murray has prepared a draft checklist of the plants of North Uist, and has made all these records available to us. Both Mrs. Murray and Mrs. J. W. Clark have recorded extensively in North Uist. Mrs. Clark and Mr. I. MacDonald have worked very enthusiastically to provide a list of Gaelic plant names, and we gratefully acknowledge the financial assistance of the Gaelic Books Council in the preparation of this publication. We also thank all those who took part in our field excursions: Arthur Chater, Mary Chorley, John and Margaret Cannon, Joanna Robertson, Ewen Cameron and Jim Bevan.

# The Geography of the Outer Hebrides

PROFESSOR W. RITCHIE

*Department of Geography; University of Aberdeen*

The 200 km long island chain of the Outer Hebrides has 119 named islands of which only 16 are now permanently inhabited (Boyd, 1979b). The archipelago contains a unique series of landscapes that are produced, in essence, by a combination of peripheral position and geological development. Exposure to the Atlantic on the west gives rise to a cool, moist climate that is characterised by periods of powerful winds. The Archaean bedrock of Lewisian gneiss provides a stable basement platform upon which recent land-forming events such as glaciation, sea level changes and ubiquitous peat-development have been superimposed. On these surfaces of glacial deposits, wind blown sand, peat and exposed bedrock successive waves of settlement have cleared, burned, grazed and cultivated, while others have turned to the seas for sustenance.

The intricate spatial pattern of these islands is revealed in Figure 1 (a and b) which illustrates the dominance of high ground on the east side of the Uists and in the central area of South Lewis/North Harris, from which a local ice cap developed in Devensian times. The east-west lines of sea lochs penetrating inland from the Minch, and the straits dividing the islands from each other interrupt the arcuate sweep of the chain of islands. Equally important is the contrast between the fault-bounded east coast which plunges to deep water in the Minches and the shallow sea bed west of the islands; a submerged platform which extends to one of the most extensive areas of continental shelf in Britain. These east-west contrasts are shown in three schematic, annotated transects, through South Uist, Harris and North Lewis (Figures 2, 3 and 4). Clearly these transects are not comprehensive and omit the distinctive nature of mountainous Barra or the sandstone based Eye Peninsula near Stornoway or the many small uninhabited islands, but these transects highlight the distinction between the low Atlantic coastlands with machair-land and crofting settlement on the west, and the areas of empty mountains and the extensive peat-covered plateaux of the east and most of the interior regions.

Settlement and associated communications have two main elements; coastal crofting townships – now mainly on the west coast (but during the population peak of the 19th century settlement was much more widespread) – and small nodal towns at east coast ferry and fishing ports. Stornoway, the administrative, service, commercial and transport centre dominates the pattern of settlement and is the only multi-functional town in the region. Other local service centres are at Castlebay, Lochboisdale, Lochmaddy, Balivanich (especially as a result of its military base and airport), Leverburgh, Tarbert and Ness.

Fishing, crofting agriculture (recently boosted by massive E.E.C. investment), tourism and weaving underpin the fragile economy which is or has been augmented by local centres of enterprise such as the oil platform fabrication yard at Arnish near Stornoway (now closed), the rocket-range support base at Ballavanich, the military developments at Stornoway airfield and scattered small scale initiatives such as fish processing and, notably, salmon farming in sheltered bays and sea lochs. In fact, the Outer Hebrides, especially in the coast sea lochs, could be described as the most important area along with Shetland for salmon farming. In 1989, an N.C.C. report on Fish Farming indicated that more than 50 sites are in the Outer Hebrides with the most important concentration being in the Uig area of Lewis. In contrast, the most extensive land use is moorland sheep grazing with some West Highland type sporting estates (deer forests) especially in south Lewis and north Harris.

The distribution of *machair* land in the islands is also shown in Figure 1. In a region of bog and rock, the shell-sand based machair, dunes and sand plains provide areas of richness and fertility. Developed from a vast influx of sand that was carried onto the low rock basement by a rising sea level over the last seven or eight thousand years (Ritchie, 1985), machair with its essentially base-rich, alkaline, free-draining soils has provided a distinctive landscape element. The level sand plains appear to have been attractive to settlers throughout pre-historic and historic times (Ritchie, 1979). Although there are records of floods and sand storms, especially

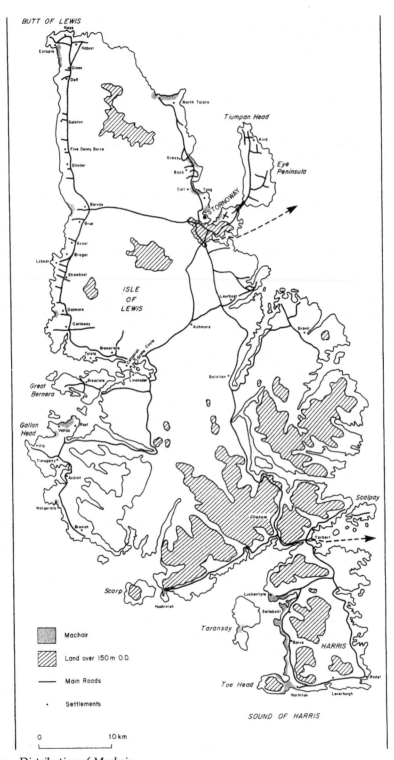

**Fig. 1a**  *Distribution of Machair*

N.

Pabbay
Ensay
Killegray
SOUND OF HARRIS
Berneray
Boreray
Ard a' Mhorain
Griminish Point
Vallay
Oronsay
Newton
Trumis Garry
Grenetote
Kirkibost
Balmartin
Tigharry
Aird an Runair
Hougharry
Balranald
Dublin
NORTH UIST
Loch Maddy
Kirkibost Island
Clachan
Monach Islands
Baleshare Island
NORTH FORD
Grimsay
Kallin
Ballivanich
Aird
Nunton
Torlum
BENBECULA
Baghan
Liniclett
SOUTH FORD
Ardivachar Point
Kilaulay
Lochdar
West Geirinish
Drimore
Stilligarry
Loch Druidibeg
Dremisdale
Howmore
SOUTH UIST
Vorran Islands
Ormaclett
Rudha Ardvule
Bornish
Mingarry
Askernish
Daliburgh
Kilpheder
Loch Boisdale
Boisdale
Garrynamonie
Ludac
Pollachar
SOUND OF ERISKAY
SOUND OF BARRA
Eriskay

Machair

Land over 150m O.D.

Main Roads

Settlements

Cliad
Traigh Mhor
Allasdale
Borve
BARRA
Tangusdale
Castle Bay

Vatersay

Sandray

0        10 km

**Fig. 1b**  *Distribution of Machair*

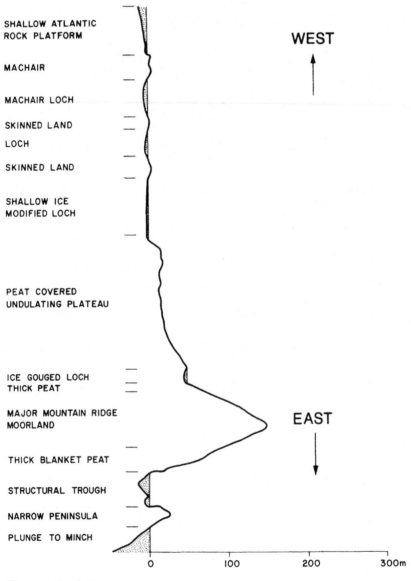

**Fig. 2**  *Transect: South Uist*

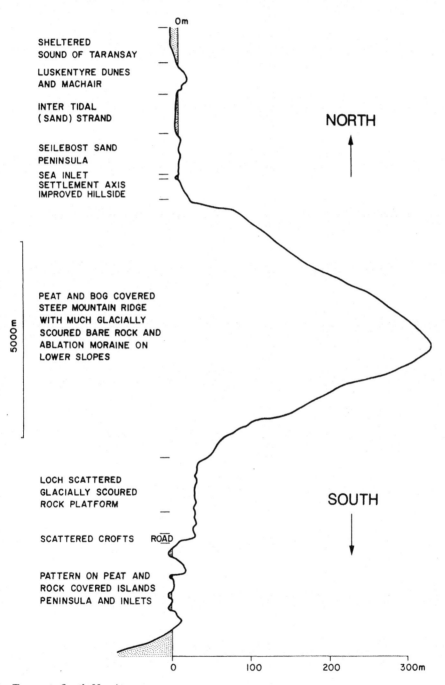

0m

SHELTERED
SOUND OF TARANSAY

LUSKENTYRE DUNES
AND MACHAIR

INTER TIDAL
(SAND) STRAND

SEILEBOST SAND
PENINSULA

SEA INLET
SETTLEMENT AXIS
IMPROVED HILLSIDE

NORTH

5000m

PEAT AND BOG COVERED
STEEP MOUNTAIN RIDGE
WITH MUCH GLACIALLY
SCOURED BARE ROCK AND
ABLATION MORAINE ON
LOWER SLOPES

LOCH SCATTERED
GLACIALLY SCOURED
ROCK PLATFORM

SOUTH

SCATTERED CROFTS    ROAD

PATTERN ON PEAT AND
ROCK COVERED ISLANDS
PENINSULA AND INLETS

0          100          200          300m

**Fig. 3**   *Transect: South Harris*

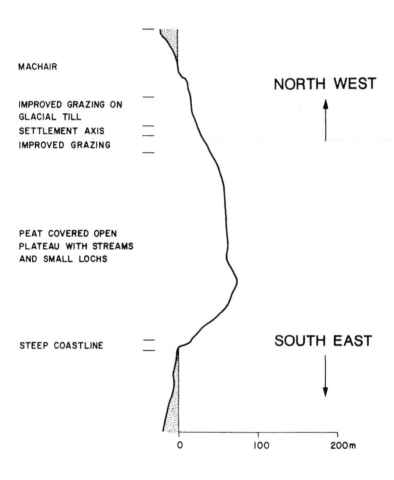

**Fig. 4**  *Transect: North of Lewis*

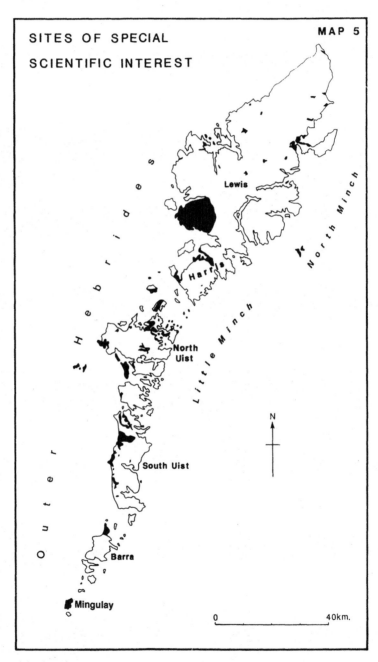

**Fig. 5** *Sites of special scientific interest*

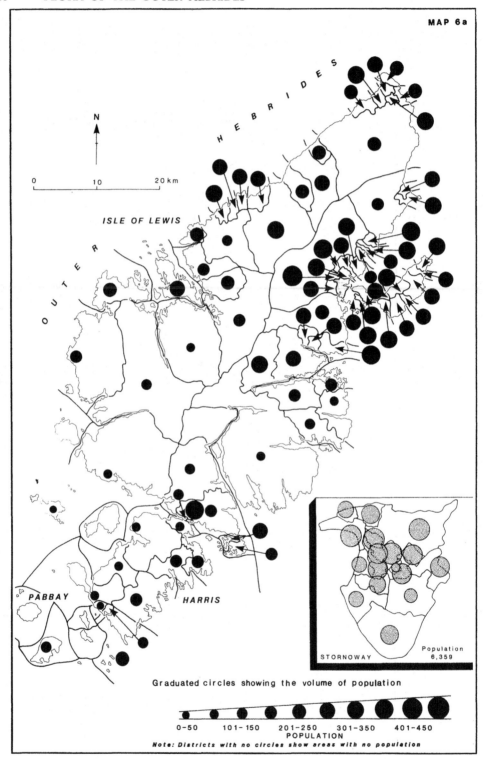

**Fig. 6**  *Population census 1981, represented by the enumeration districts*

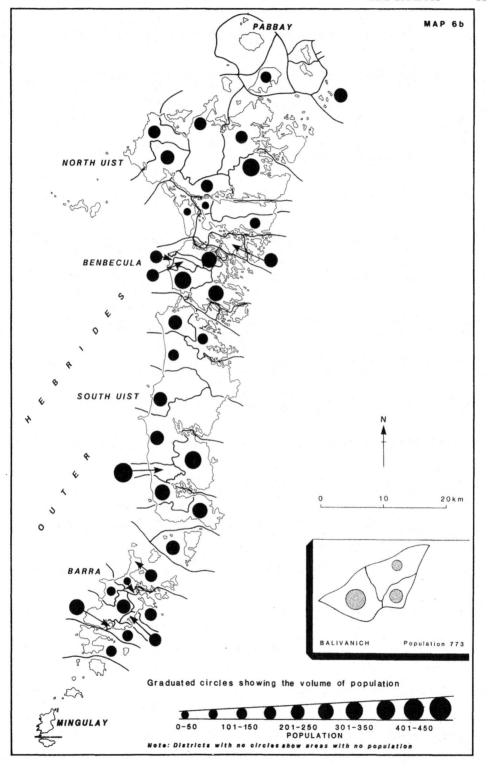

Graduated circles showing the volume of population

BALIVANICH          Population 773

0-50    101-150    201-250    301-350    401-450
POPULATION
Note: Districts with no circles show areas with no population

in the eighteenth and nineteenth centuries, machair provides both arable and grazing land and it is the focus of most Hebridean crofting agriculture (Caird, 1979). From an ecological point of view the maritime, calcareous dune and pasture areas provide zones of exceptional richness and diversity; a diversity that is enhanced by the system of non-intensive agriculture (Roberts, Kerr and Seaton, 1959) and the close juxtaposition of moorland, loch and marshland habitats. As an index of this environmental richness, Figure 5 shows the distribution of Sites of Special Scientific Interest in these islands (see appendix); a land designation that is about 10 per cent of the total area.

The distribution of machair (Figure 1, a and b) may be used as a means of identifying differences between island groups. In Barra and adjacent islands, machair is widespread but occurs as discrete bayhead units, although Eoligarry machair is a distinctive peninsula joining the northern part of Barra to the central massif. In the Uists, the machair is almost continuous along the west and north coasts, and here the area of machair land is almost ten per cent of the total land area. In Harris, there are large areas of machair in the southwest around the inner sheltered sea areas collectively known as the Sound of Taransay. In Lewis, machair is a small percentage of the total area and it occurs in a variety of forms; small bayhead units, larger east coast elements north of Stornoway, complex beaches and strands in the southwest at Uig and semi-continuous expanses between Barvas and Ness on the northwest coast.

There are other geographical differences between the islands. In Lewis, the familiar Lewisian bedrock is replaced north of Stornoway by sandstones and conglomerates. Unlike the islands to the south, Lewis and north Harris had a large local ice cap (von Weymarn, 1974, Peacock 1984). Shore platforms and possibly raised beach elements are also found in Lewis. Nevertheless, the overall impression in all the islands is of coastline submergence; a process that appears to have occurred over the last 8,000 years or so with the amount of sea level rise being estimated as approximately 5 m. Perhaps the most striking difference between the northern and southern islands is in the distribution of mountains where the high series of peaks (averaging 550 m above sea level) of north Harris and south Lewis have no equivalent elsewhere. In the Uists, the distinctive east coast barrier of mountainous ridges (rarely more than 300 m high), reflects geological structure, and is the most striking feature of Long Island topography. Similarly, the broad, low, peat-covered plateaux of central and north Lewis is repeated only in central North Uist where it resembles similar extensive, windswept and empty surfaces in mainland Sutherland and Caithness. Geologically recent submergence has also produced a series of islands, some inhabited e.g. Eriskay and Berneray, and some now empty and abandoned e.g. Fuday, the Monach Islands and Pabbay. This submergence has also produced the myriad skerries and islands of the inter-island straits, notably between North Uist and South Harris. On land, the analogue to the skerries are the countless ice-scoured and structural lochs and lochans, some of which are tidal. These provide an intricate pattern of interconnected and convoluted water bodies which have considerable ecological interest and value.

Except where crofters and their antecedents have skinned-off the peat for fuel, deep, treeless blanket bog (*blackland*) covers all the low ground of these islands. Near some crofts however some areas have been improved by liming and surface seeding and elsewhere grazing, drainage and muirburn have altered the natural vegetation cover. Land that is cultivated or grazed intensively near the croft is often referred to as *inbye* to distinguish it from common grazings and more distant fields. For thousands of years, poor drainage and an excess of rainfall over evapo-transpiration has favoured sphagnum peat growth. Peat remains a vital island fuel but little if any is exported and there is no industrial use such as whisky distilling. Peat moorland, especially at low altitude, was used in the past for cultivation, normally potatoes but sometimes cereals. The technique used was lazy-bed cultivation which is a form of hand cultivation. A few patches of such use can still be seen today, especially on small islands, but the practice has almost disappeared. Essentially, two parallel trenches are made and the turf is folded over towards the centre. The total strip is usually over one metre wide. Seaweed or other dressing might be applied in this central raised area where the crop is planted, and the trenches provide drainage. Good crops were raised by this labour-intensive form of cultivation. Literally thousands of patches of old lazy-beds can be seen throughout the Hebrides today as a clear testimony to land pressure, especially in the 19th century.

The two main part-time occupations in the crofting communities are tweed weaving with the managerial, marketing and assembly units being in Stornoway, and various types of fishing.

Fishing is vital to the economy, especially in Barra, Eriskay and different parts of Lewis. Minch and Atlantic fisheries are augmented by onshore and coastal fishing for valuable shellfish, including crabs and lobsters. Recently, there has been a rapid increase in fish farming, especially salmon, in the sea lochs of Lewis and Harris and elsewhere. Of great local importance are military establishments, especially the garrison in Benbecula which services the NATO rocket and missile firing range on the northwest coast of South Uist. Tourism, served by some large hotels but mainly bed and breakfast facilities in many croft houses, is of considerable significance. Car ferries to Castlebay, Lochboisdale, Lochmaddy, Tarbert and Stornoway link the islands with the mainland at Ullapool and Oban, and Uig on Skye. There are also small inter-island car ferries including the recently opened routes to Eriskay and Berneray. There are scheduled flights between the islands, and daily air services to Glasgow and Inverness. These modern developments contrast with the situation earlier in this century when for example there was no bridge between Benbecula and North and South Uist, and crossings had to be made across the beach at low tide.

It is often said that the best index of the geography of an area is the distribution of population in space and time. The Outer Hebrides is no exception and Figure 6 is a compilation from the 1981 census data, and it shows the low population density and the extent of areas that are effectively uninhabited, in response to overwhelming environmental and topographical diffi-culties. Census returns also show a steady decline from a late 19th century peak where arguably there was gross overpopulation, a situation that was remedied by drastic reorganisation of tenure and agricultural use and emigration, (Caird, 1979). The 20th century has seen a steady decline to the present total of 31,801 (1981 census) with particular areas showing local reversals e.g. Benbecula with the advent of the rocket range in South Uist. Nevertheless, the level of population is now approaching a dangerously low level in some areas and this is compounded by the age structure with a relative lack of young adults in the population.

Similarly, the economic base is fragile and overdependent on specific local and possibly transient opportunities e.g. military bases, industrial and commercial initiatives e.g. the oil platform fabrication yard at Arnish, and some Highlands and Islands Development Board sponsored enterprises e.g. fish processing factories. The tourist industry is also highly seasonal and subjected to fluctuating cycles. In a sense, history is repeating itself in that the sequence of development in the past has also contained local 'boom' elements such as kelp gathering in the 19th century or the bulb growing schemes in the 1960's, both of which foundered in response to changes in external economic imperatives, although at the outset there were high hopes that unique local resources i.e. seaweed and disease-free sandy soils, would provide enduring employment opportunities.

The geography of the Outer Hebrides is thus a fascinating product of the interplay of physical and human factors with powerful historical undertones. Peripheral position, the friction of distance, the lack of natural resources and an absence of low cultivable ground combine to provide a restricted and disadvantaged economic base. Celtic cultural and linguistic traditions, the difficulties of crofting (which is most effective on a part-time basis) and, here and there, implants of modern industry, sophisticated communications and strategic installations charac-terise the social and economic geography of these islands. In these respects, the islands share the common attributes and problems of regions on the distant fringes of densely populated urban-centered economic and social national structures. Overall, there is an abiding impression of emptiness, of a natural pace and rhythm and of a traditional way of life that has all but disappeared from superficially similar island communities elsewhere in the Highlands and Islands of Scotland.

# The Geology of the Outer Hebrides

## C. D. GRIBBLE

*Department of Geology and Applied Geology; University of Glasgow, Glasgow G12 8QQ*

### Introduction

MacCulloch (1819), and later, Murchison and Geikie (1861), provided the first accounts of the geology of Lewis and Harris and the other islands which constitute the north east-south west chain of islands called the Outer Hebrides, which lies some 70 km west of the northern Scottish mainland. In 1923, Jehu and Craig produced the first detailed account of the geology of this region, and followed it up with further accounts between 1925 and 1934. Thereafter, several research papers were published on the Lewisian complex, including Dearnley (1962), Myers (1970, 1971), Coward (1972, 1973) and Coward & Graham (1973); and other papers have been concerned with the geology of the igneous complex of south Harris, and in particular, the anorthosite intrusion cropping out in the south east (Davidson 1943; Dearnley 1963). A new geological map (The Outer Hebrides ) on a scale of 1:100000 was published by the British Geological Survey (BGS) in 1981, and a detailed report on the entire region is in an advanced state of preparation. The most recent published works on the geology of the islands as a whole are by Watson (1977) and Smith & Fettes (1979).

### Geological history

The Outer Hebrides are composed primarily of metamorphic gneisses and igneous rocks of Precambrian age, apart from a small area near the town of Stornoway in Lewis, which has rocks of either upper Palaeozoic or lower Mesozoic age resting unconformably on the Precambrian rocks. The Precambrian metamorphic rocks are collectively known as Lewisian, the name being obtained from the Isle of Lewis where they were first described and recognised. The Lewisian rocks formed during the late Archaean – early Proterozoic eras, roughly indicating a time span of from c. 2800 m.y. to c. 1600 m.y. This length of time can be divided into two events in Scotland, originally recognised on the N.W. Scottish mainland: namely an earlier Scourian and a later Laxfordian, the boundary between the two events being dated at approximately c. 2200 m.y. A great thrust, the Outer Hebrides Thrust or Fault, terminated the Laxfordian sequence of events, and this thrust can be recognised at the present day as a feature on or near the eastern seaboard of the entire Outer Hebrides chain of islands (see Fig. 7).

### Rock types

The Lewisian gneiss.

This comprises either quartzo-feldspathic gneiss or hornblende gneiss, often with intercalated amphibolite sheets. The quartzo-feldspathic gneiss contains quartz, feldspar and hornblende as the principal minerals, variable though small amounts of biotite, and with iron ores and apatite as the common accessory minerals present. In general, the gneiss is quartz-rich but variable in character, and with hornblende patches and acid veins occasionally present. The hornblende gneiss is similar in character and mineralogy, but hornblende is invariably the most abundant mineral present. The amphibolite bands and lenses are more or less biminerallic, with hornblende by far the most abundant mineral, often occupying more than 75 per cent of the rock, and with feldspar as the other mineral present. The grain size varies from coarse to fine grained, and the rock has a banded appearance. The textures and general characters of the gneisses suggest that these rocks have been metamorphosed to high grades such that they are migmatised (melted) in places. Although it is difficult to determine what was the original material from which these rocks were derived, the general consensus of opinion is that the original material was of igneous origin, with the old term orthogneiss being used to describe these rocks in the past.

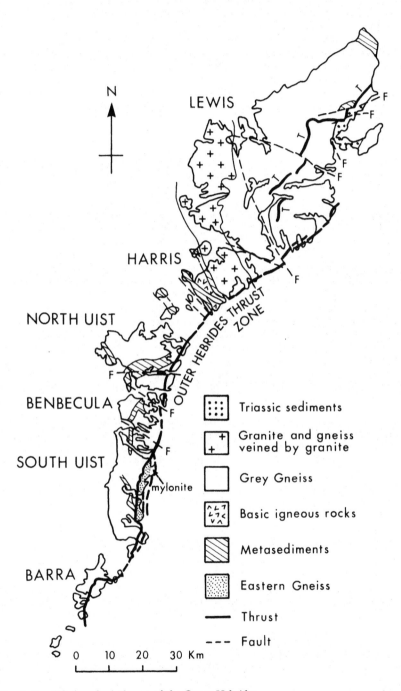

**Fig. 7**  *A simplified geological map of the Outer Hebrides*

As well as the Lewisian orthogneisses described above, several outcrops of metamorphic rocks occur which were undoubtedly derived from sedimentary rocks, and the old term paragneiss has been used to describe these, although this has now been superseded by the simpler term metasediments. Metasedimentary rocks are common in south Harris where they occur on a large scale in well-defined belts extending for considerable distances; up to 15 km in places. The metasediments include calcareous and graphite-bearing gneisses, quartzites, various marble bands, and dark-coloured garnet-bearing schists. In general these rocks have been subjected to high grade metamorphism, have a variable grain size, and occur in north west-south east trending belts, although they have also been recognised as small discontinuous lenses, a few metres in length, intercalated with other gneissic rocks. The geological map, Fig. 7, shows the distribution of the main Lewisian types. The metamorphic gneisses throughout the Hebrides have been eroded to a gentle, undulating landscape as can be seen in central Lewis.

### The igneous rocks

Igneous rocks occurring in the Outer Hebrides can be ascribed simply to one of three main groups: a) the igneous complex of South Harris, including the eastern Basic Gneiss of South Uist; b) the granites and pegmatites; and c) the basic dykes.

a)    Intrusive igneous rocks occur in south Harris and western Lewis particularly, as well as in a few other localities, of which eastern South Uist is the most important. In South Harris a complex of igneous rocks, ranging from basic to acid types, has been intruded into the meta-sediments there. The igneous complex is elongated in a north west-south east direction, and includes norite, anorthosite, tonalite and diorite as the main rock types present, all of which have been metamorphosed since their intrusion and are termed meta-igneous rocks by the British Geological Survey in the published map, with terms metanorite, meta-anorthosite, etc. being used. The igneous complex is composed for the most part of very resistant rocks, and less erosion has taken place here than with the other rock types, so that this region contains some of the highest hills in the Outer Hebrides, such as Roneval, Bleaval and Chaipaval. Other high hills in the Outer Hebrides are found where the igneous rocks of eastern South Uist crop out, and include Ben More and Hecla. The igneous complexes described have been dated at c. 2200 m.y. All the rocks described so far, that is the Lewisian gneisses and the igneous intrusions, were subjected to several phases of metamorphism and deformation, so that the mineralogy and fabric of each of these rock types are a result of its complex geological history.

b)    At the end of the Laxfordian events affecting the Lewisian rocks a suite of granite intrusions and accompanying pegmatite dykes and veins were emplaced c. 1750 m.y. ago. Their greatest development is in west and central Harris and western Lewis. A central body of granite trending NNW – SSE is flanked by a marginal belt of veined gneisses, the veins dying out away from the central body of granite.
    The granites throughout the Islands tend to be coarse grained and grey or pink in colour, and with the accompanying pegmatites very coarse grained and pink or reddish in colour, similar to the typical Laxfordian pegmatites on the Scottish mainland at Loch Laxford. Erosion of the granites gives rise to a barren, rolling landscape such as is seen in the Uig Hills of western Lewis, and steep, inaccessible sea cliffs such as are present in western Harris.

c)    Several phases of dyke emplacement took place: 1) an early group which were intruded between the Scourie and Laxfordian events and which comprise mainly basic types, particularly metadolerites. The amount of deformation these dykes have suffered has been used to unravel the main structural differences between the Scourie and Laxfordian events. To put it simply, structures cut by the dykes must be Scourian, whereas structures affecting the dykes must be Laxfordian.
    2) Several suites of basic, unmetamorphosed dykes have been emplaced into the rocks of the Outer Hebrides. These include Permo-Carboniferous quartz dolerites occurring as a swarm of east-west trending dykes in North Barra and South Uist which are cut by slightly later thin monchiquite dykes; and much later north west-south east trending Tertiary dykes found in a belt across southern Lewis and Harris.

The sedimentary rocks

Sedimentary rocks, unaffected by metamorphism, crop out in an area from Stornoway eastwards to the Eye Peninsula and northwards along the coast to Vatisker Point. The rocks primarily consist of unfossiliferous conglomerates the age of which is not known with certainty, other than being post-Caledonian, but they are considered to be of probable Trias age. The outcrop pattern near the coast shows offsetting by a couple of small east-west trending faults.

Major thrusts and faults

The most important thrust affecting the Outer Hebrides is the Outer Hebrides Thrust, a low angle thrust, dipping to the east, and cropping out along the eastern seaboard of the island chain (Francis and Sibson 1973). The thrust is shown on the map, Fig.7, as also is the wide zone of rocks affected by the movements on the thrust plane. The rocks in this zone have been affected by crushing to various degrees, the most severe effects being found close to the actual thrust plane itself. The Survey describes the gneisses in this zone as: gneiss with cataclasite fabric meaning gneiss that has suffered crushing; mashed gneiss meaning gneiss that has been severely crushed; and pseudotachylite – close to the actual plane of the thrust, meaning gneiss that has been crushed so severely it is now a black, flinty, very fine grained or glassy rock. The pseudotachylite is considered to have been formed in places as molten material, by heat generated from frictional sliding along the thrust plane. In many places this molten material was then injected into the fractured rock formed by the thrusting movements, and later, when it solidified, a veinwork of pseudotachylite occurred throughout the gneiss, 'welding' it together and eventually producing a resistant rock which, in some areas, gives rise to higher hills as at Eaval in South Uist. Later, in Caledonian times, c. 500 m.y. to 400 m.y. ago, the Outer Hebrides Thrust was reactivated, and the movements produced more crushing along the thrust plane, with the production of mylonites, which are pale coloured, very fine grained, schist-like bands of crush material occurring at irregular intervals along the line of the thrust.

As well as the crushing effects described above, many of the rocks in the thrust zone suffered hydrothermal alteration. This is particularly evident in the igneous complex in south Harris, where, for example, the anorthosite – a wedge-shaped outcrop of rock trending eastwards from Roneval hill – is almost completely altered at Lingara Bay in the east, from a pale purple, feldspathic rock to a dense, white rock, composed to a large extent of epidote group minerals. This type of alteration is called saussuritisation, and other similar types of hydrothermal alteration can be recognised in the other igneous rocks of the igneous complex of south Harris where they crop out in the thrust zone.

In South Uist, a valley called Allt Volagir is considered to lie on a north – south trending fault, which may represent an off-set of the Outer Hebrides Thrust.

Numerous faults affect the rocks of the Outer Hebrides, the most important of which are those trending north west-south east, and east- west. The most important north west-south east faults include one from Loch Seaforth to Loch Roag in Lewis, and two inferred faults between North Uist and Benbecula, and Benbecula and South Uist. Two important east-west faults which occur are from Loch Leurbost to Loch Roag in Lewis, and from East Loch Tarbert to West Loch Tarbert in Harris. The Outer Hebrides Thrust, and all the important faults mentioned above can be found in the geological map Fig. 7.

Topography and overburden

In summary, the metamorphic gneisses and related rocks give the Outer Hebrides a subdued and undulating landscape, with occasional hilly areas which represent either igneous complexes as in south Harris and eastern South Uist, or 'welded' gneisses in the vicinity of the thrust as occurs in North Uist. North west-south east and east-west trending faults have also affected the landscape with many of the sea inlets and valleys having this orientation. Relatively recent glaciation has stripped off any residual cover the rocks may have possessed, leaving them barren, particularly where granitic rocks crop out as in north Harris and western Lewis. Glaciers have also gouged out the weaker rock along the lines of ancient faults, and deepened the existing valleys there. This glaciation has given the Outer Hebrides a complex relief of rock knobs, and small, peat- or water-filled hollows, and intersecting linear grooves or valleys often occupied by lochs and lochans. In some places the glacial deposits were reworked by small glacial streams and rivers leading to occasional, small deposits of fluvio-glacial sands and gravels which are often used throughout these islands as sources of aggregate for building purposes. Near the shore, old raised-beach deposits can be observed throughout the region, representing changes in sea levels at about the same time as the glacial episode.

# Geomorphology and Soils of the Outer Hebrides

## G. HUDSON

*Macaulay Land Use Research Institute, Craigiebuckler, Aberdeen*

The arcuate island chain of the Outer Hebrides is underlain by rocks which are amongst the oldest in Britain, principally comprising Lewisian gneisses with some granites and younger sedimentary rocks. The soils, however, are of relatively recent origin, having formed since the end of the glacial epoch when the pre-existing soils and regolith were destroyed. The glacial drifts on which many soils are developed are widespread in Lewis and occur sporadically in Harris, North Uist, Benbecula and South Uist. Some soils however are on yet younger materials, the peats and windblown shelly sands which formed subsequent to the glaciations. The principal factors influencing the processes of soil formation in the Outer Hebrides are climate, parent materials, topography and vegetation, and the distribution of soil types in the islands reflects the variations in these environmental factors.

### Soil formation

In the Tertiary period, weathering produced a deep regolith (Fitzpatrick, 1963), the remnants of which exist in the Outer Hebrides at a few sites protected from severe glacial erosion (Glentworth, 1979). The general form of the landscape, low-lying in North Uist, South Uist and Benbecula, gently undulating in Lewis and mountainous in Harris, was established well before the Pleistocene glaciations, when the islands were covered by ice, whose movements were affected by radial outflow from a snow accumulation centre on Harris. Mainland ice flowing from east to west at the time of the glacial maximum was diverted northwards around Harris, whereas the smaller, isolated hills on North Uist, Benbecula and South Uist provided little restriction to the regional movements. Intense glacial erosion exploiting faults and other lines of structural weakness removed the weathered regolith to form the distinctive topography known on North Uist as 'knock and lochan'. 'Crag and tail' features, having lodgement till on protected lee slopes of rock knolls, occur sporadically, and in Lewis till was deposited to form an extensive plain. As the main ice-sheet receded, englacial and supraglacial materials were left as moundy morainic drifts in North Uist, Benbecula and South Uist.

Following the ablation of the major ice-sheets, there was a resurgence of valley glaciation in Harris and locally in South Uist, during which bouldery morainic drifts were deposited, principally in valleys. At this time also the surrounding country was exposed to periglacial activity. The distribution of these and other parent materials is shown on Figure 8. Sands and gravels deposited by water flowing from the melting glaciers are restricted to a few localities and there are no major raised beach deposits resulting from sea level changes relative to the land surface. Rather more recently, plains of windblown shelly sand have formed extensively along the west coasts of North Uist, Benbecula and South Uist and in localised bays around Harris and Lewis. The machair and dune systems developed contemporaneously with the extensive peat cover which dominates the landscapes, particularly in Lewis and North Uist.

The cool wet climate (Birse, 1971) of the islands affected soil formation in two ways. The low potential water deficit has resulted in many waterlogged soils having subsoils with anaerobic, reducing conditions which restrict plant rooting depths. Secondly, the waterlogged soils remain cold in spring, and organic matter resulting from vigorous plant growth breaks down only slowly and peat has accumulated widely in the islands. The machairs, however, are coarse-textured, readily permeable and base-rich, and organic matter breakdown under the aerobic conditions and good nutrient supply in the calcareous soils is more rapid.

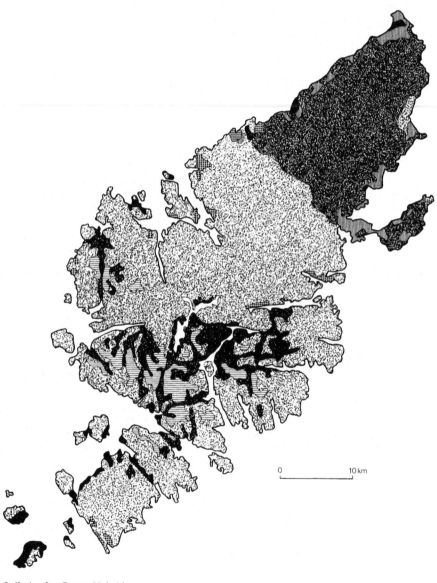

**Fig. 8**   *Soils in the Outer Hebrides*

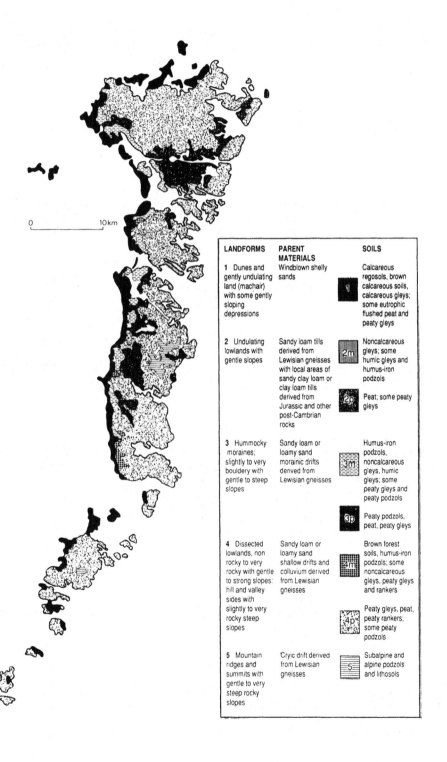

| LANDFORMS | PARENT MATERIALS | | SOILS |
|---|---|---|---|
| 1 Dunes and gently undulating land (machair) with some gently sloping depressions | Windblown shelly sands | | Calcareous regosols, brown calcareous soils, calcareous gleys; some eutrophic flushed peat and peaty gleys |
| 2 Undulating lowlands with gentle slopes | Sandy loam tills derived from Lewisian gneisses with local areas of sandy clay loam or clay loam tills derived from Jurassic and other post-Cambrian rocks | | Noncalcareous gleys; some humic gleys and humus-iron podzols |
| | | | Peat; some peaty gleys |
| 3 Hummocky moraines; slightly to very bouldery with gentle to steep slopes | Sandy loam or loamy sand morainic drifts derived from Lewisian gneisses | | Humus-iron podzols, noncalcareous gleys, humic gleys; some peaty gleys and peaty podzols |
| | | | Peaty podzols, peat, peaty gleys |
| 4 Dissected lowlands, non rocky to very rocky with gentle to strong slopes: hill and valley sides with slightly to very rocky steep slopes | Sandy loam or loamy sand shallow drifts and colluvium derived from Lewisian gneisses | | Brown forest soils, humus-iron podzols; some noncalcareous gleys, peaty gleys and rankers |
| | | | Peaty gleys, peat, peaty rankers; some peaty podzols |
| 5 Mountain ridges and summits with gentle to very steep rocky slopes | Cryic drift derived from Lewisian gneisses | | Subalpine and alpine podzols and lithosols |

*Table 1.  Major soil sub-groups, their distribution and extent in the Outer Hebrides*

| Division | Major soil group | Major soil sub-group | *Extent | Distribution |
|---|---|---|---|---|
| Immature soils | Lithosols | Undifferentiated lithosols | E | W |
| | Regosols | Calcareous regosols | M | W |
| | Alluvial soils | Saline alluvial soils | Cf | L |
| | | Mineral alluvial soils | Cf | L |
| | | Peaty alluvial soils | Cf | C |
| | Rankers | Brown rankers | Cf | L |
| | | Peaty rankers | E | W |
| Non-leached soils | Calcareous soils | Brown calcareous soils | M | W |
| Leached soils | Brown earths | Brown forest soils | Cf | L |
| | Podzols | Humus-iron podzols | M | C |
| | | Peaty podzols | M | C |
| | | Subalpine podzols | Cf | C |
| | | Alpine podzols | Cf | L |
| Gleys | Surface-water gleys | Non-calcareous gleys | M | C |
| | | Humic gleys | Cf | C |
| | | Peaty gleys | E | W |
| | Ground-water gleys | Calcareous gleys | M | W |
| | | Peaty gleys | Cf | W |
| Organic soils | Peat | Eutrophic flushed peat | Cf | W |
| | | Dystrophic peat | E | W |

| * Extent | Distribution |
|---|---|
| Cf  Confined | L  Local |
| M  Moderately extensive | C  Common |
| E  Extensive | W  Widespread |

### Soil landscapes

A number of soil landscapes can be recognised in the Outer Hebrides related mainly through solid geology, glacial drifts, weathering and the accumulations of shell sand, peat and alluvium. Five major landform regions related to soil parent materials (Figure 8) are described here and form convenient groupings for describing the soils. The regions are:

1. Machair and associated dune systems
2. Till-covered plain
3. Hummocky moraines
4. Rock-controlled lowlands and low hills
5. Mountains

Machair and associated dune systems

The machairs, gently undulating shell sand deposits, and their associated dune systems cover about 120 square kilometres and occupy long stretches of the western sea-board of South Uist, Benbecula and North Uist. Many bays on the west of Lewis and, more especially, Harris also have areas of these shelly sands, often confined between scenically dramatic rocky headlands.

The deposits have varying proportions of shell fragments and, according to Mather and Ritchie (1977), four-fifths of the beaches have a lime content of more than 40 per cent. The lime-rich soils have high values of some properties, especially pH, which is 7.5 to 8.0 in subsoils and 6.5 to 7.5 in topsoils. The machair landforms have been described by Mather and Ritchie (1977) and Ritchie (1979) as hilly, hillocky, undulating and plain and they also state that 87 per cent lie below 50 metres and 18 per cent below 10 metres O.D. Gneiss hills with smooth slopes facing the prevailing winds occur in some localities and blown sand has accumulated up to 100 metres on Eoligarry Hill (Barra) and to 150 metres on Pabbay (Sound of Harris). The hinterland of the machair plains is generally low-lying with slow landward accretion of sand taking place into wet hollows or lochs. A general traverse from sea to the inland edge crosses the beach, sometimes a storm beach with steep shingle banks, and rises into a dune fringe with a fragile vegetation cover of dune plant communities. The dunes protect the low lying and gently undulating machair plain behind from ingress of the sea at spring tides and in storms. Away from the dunes the machair level rises gradually before falling again towards wet hollows or lochs at the landward edge. Erosion has, in some areas, removed much of the sand lying above the water-table and the resulting equilibrium is maintained by the wet soil conditions which inhibit cultivation or erosion. There are five main subdivisions: dune systems, gently undulating land lying above the influence of the ground-water table, level land close to the water-table, eutrophic peaty wetlands and rocky terrain with windblown sand.

Firstly, active and stabilised dune systems with irregular or moundy landforms comprise dune ridges and hillocks up to 30 metres high but, more usually, 5 metres high and 10 to 20 metres broad with intervening hollows and channels. The principal soils are calcareous regosols, consisting of weakly developed A horizons overlying raw sand, often with buried A horizons indicating former land surfaces. Poorly drained calcareous ground-water gleys, in hollows between the dunes, are locally extensive in areas having dune slacks, where deflation of the sand has reached the water-table. The raw dune elements are colonised by northern marram grass dune pasture (Robertson, 1984) dominated by marram grass (*Ammophila arenaria*), with eyebright-red fescue dune pasture on the more stable soils with incipient A horizon development.

Occupying the zone behind the dune systems, on gently undulating to level land below 20 metres elevation, are freely drained, brown calcareous soils and calcareous regosols, where the water-table lies below soil profile depth. Erosion or accretion of sand by wind action is evident throughout and soil genesis is truncated by erosion or fossilised by deposition. The soils are relatively stable and have a dark grey-brown A horizon over light brownish grey or pale brown B horizons and white, shelly sand C horizons. Humus in the topsoil reduces infiltration into these soils and standing water accumulates on the surface after periods of heavy rain though such flooding is of short duration. Cultivations, with a two- to five-year rotation in many areas, attempt to leave the inverted turf fairly intact and are relatively shallow, to avoid exposing the loose subsoils to wind erosion.

Thirdly, low-lying level terrain where the water-table is close to the surface, and comprising calcareous ground-water gleys, is common near Loch Bee, Benbecula aerodrome and on Baleshare, North Uist. Moist or wet soil conditions prevail, the effects of gleying under anaerobic conditions increase with depth and the subsoils are blue or blue-grey in colour under dark greyish brown topsoils. The rate of removal of shell sands decreases as the erosion surface approaches the water-table and a level surface often remains. Locally, where the machair borders the gneiss landscapes, the lowest elevations are occupied by eutrophic peats or peaty gleys, and very poorly drained mineral soils form narrow corridors of wetlands fringing lochs or the adjoining land.

Lastly, windblown sand has accumulated up to a height of 150 metres, where rock-cored hills with gentle slopes lie close to beach or eroding dune areas which furnish a steady supply of sand. Many areas of this terrain with gneiss rock outcrops amongst the windblown sand occur from about 10 or 15 metres elevation. The gneisses have little influence on the soils apart from creating local shelter, where the windblown sand can accumulate. This land type occurs on Pabbay, Boreray and Berneray (Sound of Harris) and on the islands from Eriskay to Mingulay, and the soils mainly comprise calcareous regosols.

## Till-covered plain

Derived from gneisses, sandstones and conglomerates the extensive glacial tills in Lewis have sandy loam textures. Locally, near the Butt of Lewis, textures are clay loam where the till is formed from Jurassic rocks of the sea floor and, in the Ness district, there is sandy clay loam till. Lodgement till on slopes and 'crag and tail' landforms are developed locally on North Uist, Benbecula and South Uist. The land is gently undulating, slow to drain and the cool wet climate favours organic matter accumulation. Peat development is widespread and the soils include noncalcareous, humic and peaty gleys which have poor natural drainage and slow permeability. Shedding sites and steeper slopes, often with shallow coarse-textured drifts, have good natural drainage and carry brown forest soils or cultivated humus-iron podzols having dark grey-brown topsoils and bright ochreous subsoils. The land occurs as three district types; crofting land, peat-cutting areas and blanket peat.

The land under crofting tenure occurs around the coast and is cultivated in narrow strips interspersed with fallow areas which allow soil fertility and structure to recover. The soils are principally poorly drained noncalcareous or humic gleys, with some cultivated podzols on shedding sites. Along the west coast a narrow fringe of land is affected by salt spray and carries a close-cropped vernal squill maritime pasture (*Scilla verna – Festuca rubra*, Robertson, 1984) much favoured by grazing sheep.

At the fringe of the Lewis peat plain, peaty gleys with some peaty podzols occur adjacent to the inbye land of the crofting townships. Much of this ground has been reclaimed by man's activities, the peat cover gradually being removed for fuel to leave 'skinned land' which is now fenced and reseeded. Unfenced areas commonly have mat-grass (*Nardus stricta*) or heath-grass (*Danthonia decumbens*) as dominant species.

The bulk of this till plain carries an unrelieved blanket of peat which provides fuel to the crofters but has not been used commercially. Locally, there are dubh lochans and areas of eroding peat.

## Hummocky moraines

Morainic drift with a typical hummocky landform occupies about 10 per cent of the Outer Hebrides. This drift, formed as an ablation deposit from the wasting of the main ice-sheet, is extensive on North Uist, Benbecula and South Uist on low-lying ground. Bouldery morainic drift is also extensive in the valleys radiating from the former centres of ice accumulation in the Harris mountains. The drifts and, to some extent, the soils formed on them, differ in the two physiographic situations.

Firstly, in the lowlands the drifts comprise poorly sorted loamy sands or sands, with a variable content of stones and boulders. Below the plough layer or the A horizon in uncultivated soils there are cemented or indurated subsoils. These hard layers often have platy structure and resist easy penetration by water or plant roots. The well-drained areas on the mounds comprise brown forest soils and humus-iron podzols. The brown forest soils have a high level of organic matter in their A horizons and a brightly coloured B horizon. The humus-iron podzols developed in these readily-leached drifts have been cultivated in many localities and the profiles resemble the brown forest soils, though a sharp change from the A horizon to the brighter B horizon is often evident and iron pans are present in some soil profiles, at the top of the indurated layer. The acid reaction of these soils has almost certainly been ameliorated by the addition of shelly sand from the adjacent machairs. Soil water and surface runoff tend to accumulate in the hollows between the well-drained mounds and the soil patterns are related to these variations in hydrologic conditions. The wet hollows form receiving sites where conditions are reducing and anaerobic and the soils, principally gleys, have organic or humose surface horizons.

The nature of the deposits changes in the valley situation where loamy sand textures are still common, but many stones and boulders are present in the drift. In narrow valleys surface boulders dominate the landscape. On valley and hillsides, moundiness is less obvious and the more even surface is broken by stream gullies. Peaty podzols or peaty gleys occupy the mounds and steeper slopes. In the hummocky terrain of the valley floor the hollows, channels and gentle slopes are occupied by peat. Small alluvial fans occur on footslopes and have brown forest soils or humus-iron podzols developed in the coarse-textured stony deposits. Where exposure is severe due to wind-funnelling in cols at about 200 metres elevation, subalpine podzols occupy the crests of the moraine hummocks.

## Rock-controlled lowlands and low hills

Covering about 55 per cent of the Outer Hebrides, landscapes with the underlying rock near the surface have shallow drifts in hollows and crevices. Generally comprising a gently undulating to rugged low-lying peneplain, and local valley and hillsides and lower mountain slopes, the terrain includes non-rocky gently undulating land (similar to the till-covered landscape in Lewis), slightly to moderately rocky land with a variety of slopes, and very rocky land with rock pavements or abundant rocky knolls and ridges.

Near the machairs on North Uist, Benbecula and South Uist and amongst the morainic drifts on these islands are areas of slightly to moderately rocky undulating lowlands. The shallow, stony sandy loam till carries a wide range of soil types, including brown forest soils on well-drained knolls, and noncalcareous or peaty gleys in the hollows. Crofting townships occupy this land which is largely fenced and comprises improved grasslands.

Peaty gleys and peat are developed on the shallow stony tills widespread on hills near Clettraval (133 metres) and Marrival (303 metres) in North Uist, in the central area of Benbecula and, sporadically, north of the Harris mountains. Organic surface horizons are often thicker than 50 centimetres, and peat-cutting is practised in easily accessible areas around lochs and roads. Peaty gleys occupy slopes, mounds and ridges.

Moderately rocky land having peaty soils and comprising about 25 per cent of the Outer Hebrides is especially extensive on South Uist, eastern Benbecula, north of the Harris mountains and on the metasedimentary rocks in South Harris. Crag and tail features occur throughout, but are especially frequent north of Harris. Peaty gleys and peat are codominant soils on this land, with peaty rankers occupying rock knolls where glacial erosion has removed all drifts.

On the western seaboard of Lewis and Harris from Great Bernera to South Harris, there are extensive areas of very rocky land comprising dissected lowlands and hills with rock morphology ranging from crenellated to pavement. Drift is largely absent in this terrain, where glacial erosion was severe and the very shallow soils are mainly peaty lithosols and peaty rankers with some peaty gleys or peat in hollows.

Valley and hillsides with steep, slightly to very rocky slopes and having stony colluvial drifts are common in the hilly country on the eastern side of North Uist and South Uist and in the mountainous terrain of Harris. The soils are mainly peaty gleys and peaty rankers with some peaty podzols on drier sites. Surface runoff is rapid as is leaching from mineral horizons, and locally, in channels, there are flushed soils. The few crags on the steepest ground have associated short scree slopes which, in the well-drained conditions have peaty rankers with well-structured organic horizons.

## Mountains

Comprising similar terrain to the previous category, but confined to the high ground above 300 to 350 metres this land includes the summits, ridge crests, plateaux, cols and upper slopes of mountains in Harris and on South Uist. Intense rock comminution by repeated freezing and thawing has formed a stony drift which shows evidence of vertical and lateral movements of both stones and fine earth material by frost-sorting. Subalpine soils generally occupy the elevation range 300 to 600 metres in the orohemiarctic climate zone of Birse (1971). The land is gently to steeply sloping and rocky on ridge crests, plateaux, in cols and on mid-slopes of mountains. Peaty gleys and subalpine podzols are dominant, with some rankers where drift is shallow or absent. Flushed peat and peaty gleys occupy the wetter channels and hollows where snowmelt waters converge in the spring. The podzols occur on exposed knolls and on the steeper slopes.

Above 600 metres in the oroarctic climate zone, alpine podzols and gleys with lithosols and rankers occupy the mountain summits, ridge crests and upper slopes. Freeze-thaw processes disturb the soil profiles (cryoturbation) and loosen the soil fabric. Surface organic matter accumulation is reduced but humic material is translocated down the soil profile and stains the subsoils.

### Soil classification

The soil classification used here is that adopted in Scotland (Soil Survey of Scotland, 1984) for the 1:250 000 scale soil survey of the whole country during which the soils of the Outer Hebrides were mapped and described (Hudson et al, 1982). It is based principally on morphological characteristics readily identifiable in the field. The soils occurring in the Outer Hebrides are listed in Table 1, their distribution is shown on Figure 8, and brief notes of their characteristics are given below.

**IMMATURE SOILS** have indistinct or weakly developed horizons, generally restricted to surface organic horizons or A horizons resting directly on little-altered parent material or rock. They comprise principally:

**Lithosols,** which are restricted in depth and have continuous, coherent and hard rock within 10 centimetres of the surface. Only an H, O or A horizon is likely to be present above rock.

**Regosols** with a thin, weakly developed A horizon, which rests directly on unconsolidated material. The soils are formed on parent material of windblown sand.

*Calcareous regosols* contain free calcium carbonate in the parent material, shelly sand.

**Alluvial soils** are developed on recently deposited freshwater, estuarine or marine alluvium and exhibit little profile differentiation or modification to the parent material. The presence of an A or an O horizon, together with some mottling and weak structure in the subsoil, are characteristic features.

*Saline alluvial soils* have high levels of exchangeable sodium and the effects of gleying are clearly evident. The soils are developed on marine or estuarine alluvium found between the normal high-water mark and the limit of highest spring tides.

*Mineral alluvial soils* have an A horizon and the effects of gleying can be present. The soils are developed on freshwater alluvium.

*Peaty alluvial soils* characteristically have an O horizon which usually occurs at the surface but can be interbedded with freshwater alluvial sediments. The soils may have a high water-table.

**Rankers** have H, O or A surface horizons more than 10 centimetres thick which rest directly on hard noncalcareous rock or rubble derived from such rock. Incipient E and B horizons can be present.

*Brown rankers* have a brown or dark brown A horizon.

*Peaty rankers* have an O horizon up to 50 centimetres thick.

**NON-LEACHED SOILS** are characterised by the presence of free lime and have a neutral or alkaline reaction. Their lower horizons may show some gleying.

**Calcareous soils** are freely drained soils containing free calcium carbonate within the profile.

*Brown calcareous soils* contain carbonate materials in the form of rock or shell fragments. There is a gradual change between all horizons, the B horizons having a brighter colour than the A or C horizons but showing no morphological or chemical evidence of translocated sesquioxides.

**LEACHED SOILS** are characterised by a uniformly coloured B horizon, by an absence of free lime and by an acid reaction in their A and B horizons. Their lower horizons may show some gleying.

**Brown earths** have a uniformly coloured B horizon, a mull or moder humus type and a moderately acid reaction; usually each horizon merges into the one below.

*Brown forest soils* are freely drained soils having the properties of the brown earth major soil group.

**Podzols** have surface H or O horizons successively underlain by a grey, bleached E horizon and a more brightly coloured B horizon. They have a strongly acid reaction and their B horizons often contain illuviated sesquioxides of iron and aluminium and organic matter.

*Humus-iron podzols* have surface L, F and H horizons. A thin Ah horizon, not always present, overlies a pale-coloured E horizon with a low organic-matter content. The B horizons include a dark-coloured, humus-enriched, upper layer (the Bh horizon) and a bright-coloured lower layer (the Bs horizon), the latter usually containing translocated iron and aluminium. Some variation in drainage status may be found.

*Peaty podzols* have an O horizon up to 50 centimetres thick. The E horizon is generally gleyed. A horizon of humus accumulation may be present above the iron pan (Bf horizon) which is often continuous and forms a barrier to water or roots. Below the iron pan the Bs horizon is usually bright-coloured, but some variation in drainage status may be found.

*Subalpine podzols* have L, F and O horizons. Beneath the E horizon, which is darkened by organic matter, the Bh horizon is thick and very dark in colour with distinct organic coatings on small stones. The iron pan (Bf horizon) is generally weakly developed and discontinuous. Although brightly coloured, the Bs horizon is often thin and weakly developed, merging into the underlying unaltered parent material.

*Alpine podzols* have either thin L, F and H horizons with an underlying E or A horizon, or a surface A horizon with bleached sand grains. The Bh horizon is black and sometimes thick, with small stones stained with colloidal humus; the lower part of the horizon often appears less organic-stained but retains a dark reddish brown colour. The soil fabric is characteristically loose and shows the effects of freeze-thaw processes which can be intense at high altitudes.

**GLEYS** develop under conditions of intermittent or permanent waterlogging. A pale-coloured Eg horizon is often prominent in the upper mineral horizons, beneath which the horizons are grey with greenish and bluish tinges and with ochreous mottling. These colours are of secondary origin and replace those inherited from the parent material.

**Surface-water gleys** display strongly gleyed sub-surface horizons, the intensity of gleying decreasing with depth. The colour inherited from the parent material is more apparent as gley phenomena decrease.

*Noncalcareous gleys* have no free calcium carbonate in the upper horizons of the profile. An A horizon is often underlain by an Eg horizon which may be well defined in semi-natural soils. The soils are often developed on parent materials of moderately fine and fine texture.

*Humic gleys* have no free calcium carbonate in the upper mineral horizons and have a dark-coloured Ahg horizon.

*Peaty gleys* have no free calcium carbonate in the upper mineral horizons of the profile. Beneath an O horizon up to 50 centimetres thick, organic staining of the Eg and Bg horizons is often present.

**Ground-water gleys** develop under the influence of a high ground-water table. The effects of gleying increase with depth and the colour inherited from the parent material is not apparent in the lower soil horizons, which are often grey or bluish grey.

*Calcareous gleys* have free calcium carbonate in the Bg horizon.

*Peaty gleys* have a surface O horizon up to 50 centimetres thick and do not contain free calcium carbonate in the upper mineral horizons. Below the Egh horizon, mineral horizons are intensely gleyed and often humus-stained.

**ORGANIC SOILS** are formed under waterlogged conditions and contain very high amounts of organic matter down to an arbitrary specified depth.

**Peat** is an organic soil which contains more than 60 per cent of organic matter and exceeds 50 centimetres in thickness. It can develop in areas of moderate to high rainfall, low mean annual temperatures and high relative humidity and under the influence of ground-water in depressions or basins.

*Eutrophic flushed peat* is flushed by seepage waters rich in mineral plant-nutrients and usually supports a wide range of plant species, including reeds, tall sedges and flush alderwood where not reclaimed.

*Dystrophic peat* is not affected by flushing. The peat supports plant communities dominated by dwarf shrubs, usually heather.

# Climate and Vegetation of the Outer Hebrides

I.S. ANGUS

*Nature Conservancy Council, 17 Francis Street, Stornoway, Isle of Lewis*

## Introduction

Plant life on islands, as elsewhere, is most affected by climate, substrate, other organisms, and geographic situation, the last in relation to island size and situation in relation to the source of the flora. In the Western Isles, climatic influences are almost as important as parent material in determining soil types (more important in the case of peat, the most widespread soil type), while they also exert some influence with regard to biogeography.

Meteorological measurements have been recorded at Stornoway since 1856, though the present weather station at Stornoway Airport was not established until 1942, when the Meteorological Office also began recording from Balivanich Airport, Benbecula (Manley 1979). Records have also been taken over varying periods from a range of widely spaced localities, including (sporadically) outlying stations such as St Kilda.

## Precipitation and humidity

Recorded annual precipitation ranges from just over 1000mm in parts of the Broad Bay area of Lewis to over 2400 mm on some of the higher summits of Harris. The totals for Stornoway and Balivanich are 1094 mm and 1220 mm respectively. The high ground of Harris and Uig receive proportionally more orographic rainfall, with the summits receiving about 1000 mm more rain per year than adjacent western coasts. North and north-east Lewis lies in the rain shadow of these hills; the Uist machairs are too low-lying to receive much orographic precipitation, while St Kilda, with a summit at 426 m, is sufficiently high to develop highly localised weather conditions, with increased rainfall, humidity, cloud cover and wind exposure (Campbell 1974).

May and June are the driest months, together accounting for only about 10% of the annual rainfall, while more than 12% falls in December alone. Days with measurable rainfall (>0.2mm) are fairly evenly spread over the year, with monthly averages ranging from 17 to 26, and a total of 263 days (Manley 1979), of which 200 are 'wet', with more than 1 mm recorded. On higher ground, the number of such wet days rises to over 220 days per year. Average relative humidity is high throughout the year, ranging from 80 to 88% saturation, with an annual mean of 85%.

Though snow is observed falling on 47 days a year at Stornoway, it is lying (at 0900 hrs) on only 11 of these days (5 in the Uists). Late snow beds are rare in the hills, though limited areas may survive on screes in north-facing corries which receive no winter sun and relatively little summer sun. Snow may also lie later in block screes (which also tend to be located below north-facing slopes because they were mostly formed by freeze-thaw action).

The ecological significance of these precipitation figures in terms of soil moisture may be judged by relating them to the amount of moisture lost through evaporation and transpiration over a given period. If rainfall exceeds potential evapo-transpiration losses there is a Potential Water Surplus, and a Potential Water Deficit if the converse is the case (Green 1964). Most of Lewis and Harris and the uplands of South Uist have a very low Potential Water Deficit, as precipitation equals or even exceeds evapo-transpiration in all months of the year, and the uplands of southern Lewis, Harris and South Uist are so moist that even in summer (April-September) there is a PWS of over 500 mm. This is related to the fact that rainfall increases with altitude, while evapo-transpiration decreases (Birse & Dry 1970).

A long history of water surpluses over much of the area of the Outer Hebrides has led to severe leaching of many Hebridean soils, and to the formation of peaty podzols, gleys and peaty gleys and ultimately peat itself (Hudson *et al.* 1982).

*Sunshine*

Average daily sunshine at Stornoway is 3.40 hours, and 3.89 hours at Benbecula (Manley 1979). The slightly lower figure for Stornoway may be caused by a greater tendency for cloud formation over the land to the west and south-west (the sources of the prevailing winds), while exposed western coasts tend to have good sunshine records (at sea level)(Green & Harding 1983).

That the sunniest months are May and June is related to the tendency for anticyclones to develop in the Atlantic during these months bringing fine, clear weather to the islands, sometimes for long periods (Manley 1948).

The relatively high latitude of the islands means that days in midsummer are long, with very long twilights, with correspondingly short days in winter. The latitude also reduces the angle at which the sun's rays strike the earth, lengthening shadows and increasing the filtering effect of atmospheric haze (Green 1964). Steep-sided, north facing corries such as Coire Roineabhail in South Harris may be in perpetual shadow in winter and receive very little sunshine even in summer. Even where direct sunlight is received for much of the day, its heating effect is severely restricted by high winds and high humidity (Green & Harding 1983).

*Wind*

The prevailing winds originate from the south and south-west, a consequence of the frequency with which depressions (with their anti-clockwise circulation) pass to the north of the islands. The average wind speed at Stornoway is 14.4 knots (7.4m/s) (Birse & Robertson 1970) and gales are recorded on 50 days a year (Manley 1979). Indeed, it has been said that the north-west of Scotland has a higher sustained wind speed than any other inhabited part of the world (Gloyne 1968).

Birse & Robertson (1970) have classified only a few long, deep valleys of the Western Isles as 'moderately exposed', with average wind speeds below 8.5 knots (4.4m/s). Most of the land area was classified as 'very exposed', with average wind speeds of 12–15.5 knots, but only the highest summit areas were 'extremely exposed', with average wind speeds exceeding 15.5 knots (8m/s).

High average wind speeds tend to stunt the upward growth of plants and encourage the lateral growth of dwarf forms, e.g. of *Calluna vulgaris* or *Juniperus communis*, though prostrate forms of the former are not encountered as frequently as expected on exposed mountain plateaux. A *Racomitrium*-rich prostrate *Calluna* heath has, however, been noted at an altitude of only 200 m on Mullach an Reidhachd, North Harris (NB090143) (Brown 1986). Wind also has a profound effect on plant growth in the Western Isles in that is usually salt-laden, particularly when it has come from the west or south-west, having passed over long distances of wave-torn ocean. The salt content of the air drops off significantly with increasing distance from the sea, though local variations in topography affect this relationship. When maximum gust speed exceeds 30 knots, the salt content of the wind increases proportionally as there is more sea spray, and the spray tends to be carried further inland. (Holden 1961, Randall 1973). Studies on the Monach Isles National Nature Reserve suggest that salt-laden winds affect taller plants most, so that exposed coastal plant communities tend to consist mainly of low-growth plant forms – not just because of the stunting effect of the wind – but also because of the salt it carries (Randall 1973). The prostrate form of *Calluna* recorded from Hirta, St Kilda (McVean 1961), may be linked more to salt-burn than to altitudinal descent of the montane growth form. In many coastal locations the maritime lichen *Ramalina siliquosa* may grow in profusion far inland, while cliff-top saltings occur on many exposed western coasts, notable in Uig, Lewis. *Glaux maritima* has been reported from cliff-tops over 200 m high on Mingulay (Clark 1956). Salt-laden winds also increase the salinity of lochs close to the west coast, and may raise the nutrient content and base status of coastal blanket bogs as in western Ireland (Sparling 1967), possibly explaining the presence of plants such as *Schoenus nigricans*, which grows profusely in some coastal peatlands in the Western Isles.

In conjunction with changing Postglacial sea levels, high winds have driven large quantities of mineral sand and shell fragments inshore and inland to form machair, arguably the single most important habitat in the Western Isles for man as well as wildlife. While it is well known that machair may extend as much as 2 km inland in South Uist, less well known is the altitude to which blown shell sand exerts an ecological influence: the dunes at Luskentyre Banks (NG 0699) are up to 35 m high, while the presence of large amounts of wind-blown shell fragments is evident from the presence of charophytes in lochs on Tairaval (NB 1135) (Angus, unpublished NCC report) and at Mangersta (NB 0131) (Biagi et al 1985), both sites being more than 50 m above sea level. Currie (1988, in litt.) has also remarked on the presence of wind-blown sand high on hills at Cnoc an Fhithich (NF6504) and Ben Eoligarry (NF7007), both Barra; on Beinn Sleibhe, Berneray (NF9283); and on slopes above Traigh an Taoibh Thuath, Northton, Harris.

Wind has played its part in restricting the distribution of woodland in the Western Isles, in conjunction with deteriorating climate in the late Postglacial period (Angus 1987 – Birks, this volume) though grazing and burning of moorland have also played their part. Today most of the remaining fragments are confined to sheltered gorges or inaccessible crags or to some islands in inland lochs, the latter often displaying a wind-contoured profile.

Aspect is often important in exposed areas in relation to prevailing winds: in North Harris the rare Atlantic liverwort *Herbertus aduncus* is frequent, but in some areas is more profuse on north-facing slopes, usually around boulders where there is additional shelter. On the summit ridges *Racomitrium* is often confined to the lee side of boulders.

### Temperature

The truly maritime nature of the climate of the Western Isles is reflected in the way that air temperatures tend to be more influenced by sea temperatures than sunshine. Thus, according to mean daily temperatures, the warmest months are July and August (12.9°C) rather than the sunnier months of May and June, while the coldest are January and February (4.1°C), displaying a similar, if less marked, lag. This gives an annual range of only 8.8°C, one of the lowest ranges in Britain.

There are on average 47 days of air frost a year at Stornoway, 33 in Benbecula, and an estimated 20 in Barra. The latest and earliest dates for frost are about 1st April and 1st December respectively in the Uists, giving a particularly long frost-free period, matched in Britain only in the southern Inner Hebrides and in the Scilly and Channel Isles (Green 1964).

Solifluction terraces are present on some of the higher plateaux (Peacock 1984); solifluction hummocks and terraces arise from alternate freezing and thawing (cryoturbation). A report of plough marks behind stones on Oreval (662 m), North Harris, (A. Brown, pers. comm.) suggests that these are still active, indicating a marked contrast in frost frequency between some exposed plateaux and the coastlands of the Western Isles.

Some authors have attributed the presence of certain southern plants such as the moss *Myurium hochstetteri* to the relatively frost-free winters while Clark (1956) believed that local strains of heather were less resistant to frost, resulting in a notable loss of *Calluna* in Harris in the winter of 1946-47.

In terms of accumulated temperature (day°C above a growth threshold of 5.6°C) the warmest parts of the Western Isles are in southern South Uist and Barra. Most of the land is 'fairly warm' but of course becomes cooler with increasing altitude. Only the higher summits are 'very cold', and there are no 'extremely cold' areas with annual accumulated temperature lower than 300 day°C (Birse 1971).

It is now known that many plants grow at temperatures as low as 0°C, albeit slowly, and it is thought that higher temperatures are more important in the earlier half of the year. New figures for accumulated temperature have been calculated using 0°C as the growth threshold and covering the months January to June inclusive. The values of this 'lower quartile of accumulated temperature' are 1179 day°C at Stornoway, 1265 at Balivanich, and 1311 in Tiree (cf. Lerwick, 954; Kirkwall, 1073; Wick, 1066; Inverness, 1246; Dumfries, 1249; Cambridge, 1413; Penzance, 1658) (Bibby et al 1982).

### Oceanicity

Continental climates become increasingly maritime or 'oceanic' westwards, an effect super-imposed on the expected effect of increasing latitude. With increasing oceanicity, precipitation increases and is more uniformly distributed over the year, the number of 'wet' days increases, and humidity and wind speeds rise, while annual temperature range decreases (Poore & McVean 1957, McVean & Ratcliffe 1962). All the islands of the Western Isles are classed as 'hyperoceanic', this being the most oceanic sector of Eurasia (Birse 1971).

Ratcliffe (1968) has pointed out that lower-lying areas of the Outer Hebrides are rather drier than the mountains of these islands, but are nevertheless much wetter than low-lying areas of the eastern Highlands. The proliferation of gullies and coastal indentations gives increased humidity on a local scale.

The extreme oceanicity of the climate of the Western Isles has a profound effect on vegetation, so that montane plants grow at much lower altitudes than in more continental areas. In the Western and Northern Isles and on the north-west mainland of Scotland, it is not uncommon to find montane plants growing at or near sea level, though this phenomenon is probably better demonstrated in NW Sutherland and Shetland than in the Western Isles. Species exhibiting altitudinal descent in the Outer Hebrides include *Arctostaphylos uva-ursi, Polygonum viviparum, Oxyria digyna, Salix herbacea, Saxifraga oppositifolia, Sedum rosea,* and *Silene acaulis* (Currie 1979).

The combination of high precipitation, water surpluses, large number of wet days and low insolation experienced in highly oceanic areas is thought to increase the rate of growth and the competitive abilities of bryophytes, notably *Sphagnum* species, hypnoid mosses and *Racomitrium* (Poore & McVean 1957), all of which are very conspicuous in plant communities in the Western Isles.

West Atlantic bryophytes are particularly well represented in the upland plant communities of the Western Isles, a feature undoubtedly linked to the high number of wet days on the higher ground of southern Lewis, Harris and South Uist (Ratcliffe 1968). The Nature Conservancy Council's Upland Vegetation Survey recorded 14 patches of hepatic-rich *Calluna vulgaris* heath in North Harris, with a total area of 65 ha, the fourth highest figure recorded by the Survey in Scotland (Hobbs 1988). Liverworts in these communities include *Herbertus aduncus, Scapania gracilis, Plagiochila spinulosa, P.carringtonii* and *Mylia anomala,* as well as the filmy fern *Hymenophyllum wilsonii* (Brown 1986). The moss *Campylopus shawii* is another of the Atlantic bryophytes particularly common in wet heaths in North Harris (Ratcliffe 1968,1977b). The need for a saturated atmosphere may give rise to a particular abundance of these Atlantic bryophytes on NE- and E-facing slopes (Ratcliffe 1968), as noted above for *Herbertus aduncus.* Some Atlantic bryophytes such as *Pleurozia purpurea* (common in many blanket bogs) descend to low levels in the Western Isles.

The combination of climatic factors results in late flowering of many species in relation to their counterparts elsewhere. This is particularly noticeable on the machair, which usually retains its bare winter appearance till the end of May, and those who wish to see it at its best should wait until mid-June or even July.

*Acknowledgements*

I am grateful to Martin Ball, Andrew Currie, Mary Elliott, A. S. Maclennan and D. B. A. Thompson for their helpful comments on a draft of this chapter. I would also like to thank the Nature Conservancy Council's Upland Vegetation Survey and Andrew Currie for allowing me to quote from their unpublished reports.

# Floristic and Vegetational History of the Outer Hebrides

H. J. B. BIRKS

*Botanical Institute, University of Bergen, Allégaten 41, N–5007 Bergen, Norway.*

## Introduction

The present-day landscape of the Outer Hebrides is one of the most distinctive in the British Isles. It is characterised by being almost completely treeless and by having an almost continuous blanket of infertile acid bog and moorland dotted with innumerable lochs of all shapes and sizes and bare outcrops of the underlying acidic Lewisian gneiss. Lewis is a wilderness of bog and loch, Harris abounds in bog interspaced by bare rock, cliffs and block-fields, and boulder-strewn mountains, and the Uists are a mosaic of lochs and bogs, bordered on the west by a fertile coastal strip. There are numerous crofting townships scattered around this coastal strip, wherever areas of calcareous wind-blown shell-sand, sand dunes, and pasture (machair) occur. Such areas are common along the western coasts of Barra, S. Uist, Benbecula, and N. Uist, and are local on Harris and Lewis. Machair forms about 10% of the land area.

Away from the species-rich machair, the present-day flora and vegetation are species-poor, with only about 25 phanerogam species comprising most of the vegetation of bog, moorland, and loch. In a few sheltered rocky glens and on inaccessible cliff ledges and islands in larger lochs, dense stands of scrub of *Salix aurita* and *S. cinerea* subsp. *oleifolia* occur with scattered bushes of *Betula pubescens*, *Corylus avellana*, *Lonicera periclymenum*, *Populus tremula*, *Rosa afzeliana*, *Rubus fruticosus* agg., and *Sorbus aucuparia* (Spence, 1960; Bennett and Fossitt, 1989). In these ungrazed areas there is often a lush growth of *Luzula sylvatica*, ferns such as *Athyrium filixfemina* and *Dryopteris dilatata*, and tall-herbs such as *Angelica sylvestris* and *Filipendula ulmaria*. In a few ravines and glens in the Uists warmth-demanding, southern-atlantic cryptogams occur locally, including the Macaronesian-tropical species *Jubula hutchinsiae* and *Dryopteris aemula*. Their occurrences suggest mild sheltered conditions (Ratcliffe, 1968). The montane flora is poor, reflecting the widespread, strongly acidic bedrock, the abundance of blanket peat unsuitable for many montane species, and the limited extent of open ground at high altitude. A few basiphilous species occur locally on cliffs, including *Draba incana*, *Saxifraga oppositifolia*, *Saussurea alpina*, and *Silene acaulis*. Lochans in the S.Uist machair provide the only British localities for *Potamogeton epihydrus*, a plant with a so-called amphi-atlantic distribution centred on eastern North America and the western seaboard of Europe. Other Hebridean species with comparable distributions include *Najas flexilis* and *Spiranthes romanzoffiana*. Curiously this phytogeographical element in Scotland and Ireland consists almost entirely of aquatic or mire plants (Dahl, 1959, 1987; Perring, 1962). Amongst the bryophyte flora, there are several British endemic (e.g. *Oxystegus hibernicus*) or near-endemic species (e.g. *Campylopus shawii*, *Leptodontium recurvifolium*, *Myurium hochstetteri*). Many of the latter group exhibit spectacular disjunctions in their world range, occurring, for example, outside the British Isles only in Macaronesia, West Indies, or Queen Charlotte islands off the west coast of Canada (Ratcliffe, 1968).

The present flora, vegetation, and landscape of the Outer Hebrides thus pose several interesting and challenging questions. Did plants survive the glaciations in the Outer Hebrides, as Heslop Harrison (1948b, 1953) hypothesised for amphi-atlantic species and bryophytes such as *M. hochstetteri*? Have the flora and vegetation always been as species-poor as today, or has this poverty developed over thousands of years during the post-glacial as a result of habitat loss, soil deterioration, or land-use change? Was the mountain flora once richer, for example during the late-glacial period of about 13,500 – 10,000 years ago when deglaciation was occurring and soils were more fertile? Were trees once more widely distributed and more abundant before the onset of human interference, intensive grazing, moor burning, and blanket-bog development? If trees were once common, what were the major species – birch, hazel, rowan, oak, elm, alder, pine? How much of the present landscape is a result of human interference over thousands of years and how much is a result of natural processes of soil degradation in an extreme, oceanic environment? Was the climate once milder, and hence could southern-atlantic ferns and bryophytes have once been more widespread? Are their present localities relicts of this wider distribution?

Answers to these and related questions can be provided by palaeoecological studies involving reconstructions of the floristic, vegetational, and by inference, environmental history of the Outer Hebrides from pollen and macrofossil (seeds, fruits, wood remains, etc.) analyses of peats and loch sediments. Pollen preserved in such sediments provides a record of the local and regional flora and vegetation, whereas macrofossils reflect more local conditions. An independent chronology for these reconstructions is essential and this is provided by radiocarbon-dating of organic material preserved in the sediments.

Although at first sight, the great abundance of bogs and freshwater lochs on the Outer Hebrides suggests an abundance of suitable sites for palaeoecological studies, on closer examination several are unsuitable for detailed investigation. Almost all the bogs have been extensively peat-cut for domestic fuel. Some of the machair lochs have little or no sediment because of intensive wind fetch and wave erosion. Deposits of late-glacial age that accumulated in small closed basins within glacial drift in the valleys are now often covered by a thick mantle of peat and are difficult to locate except by extensive systematic borings. In addition many promising lochs and undisturbed bogs (e.g. Goode and Lindsay, 1979) are remote and difficult to reach with heavy coring equipment. These problems, coupled with the lack of interest amongst palaeoecologists in extreme northern and western areas until quite recently (Birks, 1988), result in comparatively little being known about the detailed floristic and vegetational history of these ecologically unique islands.

*Vegetational history before the late-glacial (pre-13,500 years before present)*

Nothing is known about the flora and vegetation of the Outer Hebrides in any previous interglacials, as no unambiguous interglacial deposits have, so far, been discovered. A possible interglacial sequence occurs at Toa Galston, northern Lewis (Sutherland and Walker, 1984). Its pollen assemblage is dominated by Gramineae, Cyperaceae, and Ericaceae, with a variety of open-ground herbs, all of which occur today on Lewis. The most interesting macrofossil is *Sphagnum imbricatum*, which today is local on Hebridean bogs. The age of the Toa Galston peat bed is not known, as it lies beyond the range of radiocarbon-dating (more than c.47,000 years before present (B.P.)).

Detailed mapping and dating of geomorphological features (Sutherland and Walker, 1984; Sutherland, 1984) indicate that part of northern Lewis was ice-free during the last glaciation and the last Scottish ice-sheet did not extend beyond the Outer Hebrides, contrary to the widespread assumption that this ice-sheet extended to the edge of the continental shelf. It is possible that the Outer Hebrides supported an independent small ice-cap (Flinn, 1978; von Weymarn, 1979). This geological evidence raises the question whether any plants survived the entire last glacial stage (*ca.* 125,000 – 13,500 B.P.) in the Outer Hebrides, as Heslop Harrison (1948b, 1953) and Dahl (1954, 1955) proposed on the basis of present-day distribution patterns. Palaeoecological studies of deposits dating from 27,000–15000 B.P. on Lewis and Hirta in the St Kilda archipelago provide important insights into the Hebridean flora and vegetation during parts of the last glacial stage.

An organic-rich silt lens at Tolsta Head, Lewis is overlain by over 2 m of glacial till. The silt is about 27,000 B.P. and its fossil assemblages suggest a herb-dominated vegetation with *Salix* (inc. *S. herbacea*) and *Juniperus* and open-ground taxa such as *Artemisia*, *Lycopodium selago*, and *Armeria maritima* (von Weymarn and Edwards, 1973; Birnie, 1983). Botanically the most interesting feature is the presence of *Koenigia islandica* pollen (Edwards, 1979), indicating that *K. islandica* grew on Lewis about 27,000 years ago. It is not known today from the Outer Hebrides, being confined in the British Isles to Skye and Mull.

An organic sand overlain by local periglacial and glacial deposits occurs on Hirta (Sutherland *et al.*, 1984). It dates from about 14,600–24,700 B.P. and contains a herb-dominated pollen assemblage with occasional grains of *Helianthemum*, a genus very rare in western Scotland today and absent from the Outer Hebrides.

These recent studies establish the presence of plants, including some arctic, arctic-alpine, and open-ground herbs on Lewis and Hirta during parts of the last glacial stage. The present geological and palaeoecological evidence suggests that there is a possibility that some taxa with northern distributions today *may* have survived the *entire* last glaciation in parts of the Outer Hebrides (cf. Dahl, 1954, 1955, 1987); however, the glacial survival of amphi-atlantic species such as *Potamogeton epihydrus* or *Najas flexilis* (as suggested by Harrison, 1948b) or bryophytes such as *Myurium hochstetteri*, *Leptodontium recurvifolium*, or *Campylopus shawii* seems unlikely in view of reconstructed July mean temperatures during the last glacial maximum on St Kilda of 4°C and mean sea-surface temperatures around Rockall of 0–2°C (Sutherland et al., 1984). The history of these and phytogeographically similar species in the Outer Hebrides remains, as elsewhere, a major phytogeographical mystery.

*Late-glacial vegetational history (13,500 – 10,000 years B.P.)*

The late-glacial period comprises the time of deglaciation after the last glacial maximum, and reflects major climatic amelioration. This trend was interrupted at about 11,000 – 10,300 B.P. by a widespread climatic deterioration that resulted in the redevelopment or expansion of corrie glaciers and small ice-caps in the Scottish Highlands, the so-called Loch Lomond stadial.

Preliminary observations on Harris (Birks, unpublished) suggest that local glaciers may have been present during the stadial in some of the valleys around Clisham, Ullaval, Tirga Mor, and Uisgnaval More. On St Kilda solifluction and the formation of pro-talus ramparts occurred at this time (Sutherland et al., 1984). No pollen analyses of undisputed late-glacial deposits from the Outer Hebrides have been published. As discussed in the introduction such deposits are difficult to locate because of extensive peat cover in the lowlands. However, the late-glacial of the Outer Hebrides is currently being studied in detail by M. J. C. Walker, J. J. Lowe, D. G. Sutherland, and associates. The results of their work will be of considerable interest, and will permit important comparisons to be made with the late-glacial flora and vegetation of Skye (Birks, 1973; Walker *et al.*, 1988) and Mull (Lowe and Walker, 1986).

Pollen spectra of early post-glacial age from Lewis (Birks and Madsen, 1979), S. Uist (Bennett and Fossitt, unpublished) and St Kilda (Walker, 1984) provide a glimpse of what the late-glacial flora and vegetation may have been like. Open species-rich grasslands with *Ranunculus*, *Plantago maritima*, *P. major*, *Thalictrum*, and scattered *Juniperus* and *Empetrum nigrum*-dominated heaths were *probably* widespread. Tall-herb and *Salix* communities with *Rumex acetosa*, *Filipendula*, and abundant ferns may have been common in wetter areas. On shallow soils and exposed sites, open communities occurred with, for example, *Lycopodium selago*, *L. alpinum*, *L. clavatum*, *Armeria maritima*, *Artemisia*, *Saxifraga oppositifolia*, *S. hypnoides*, *S. stellaris*, and *Helianthemum*. The vegetation as a whole may have resembled, physiognomically at least, modern species-poor subalpine vegetation of western Norway. All the indications are that the Hebridean flora and vegetation were more diverse than today, presumably as a result of the widespread more fertile, unleached mineral soils present immediately after deglaciation and the cessation of periglaciation. Almost all the species present then still occur today on the Outer Hebrides, for example in the mountains or on open, mildly basic soils on sea-cliffs. *Helianthemum* is extinct.

*Post-glacial vegetational history (10,000 – 0 years B.P.)*

There are only a few detailed pollen diagrams from the Outer Hebrides that cover all or part of the last 10,000 years, the so-called post-glacial. Investigations in Lewis (Erdtman, 1924), S.Uist (Heslop Harrison and Blackburn, 1946; Ritchie, 1985), Barra (Blackburn, 1948), Benbecula (Ritchie, 1966), Pabbay (Ritchie, 1985), and St Kilda (McVean, 1961) are unfortunately not sufficiently detailed to permit any useful reconstruction of the post-glacial floristic or vegeta-tional history. However, they all show very high frequencies of non-tree pollen, mainly Ericaceae and Gramineae. When calculated on an appropriate pollen sum, they suggest that there was never any extensive, continuous forest cover.

The most complete and reliably dated sequence comes from near Little Loch Roag, Lewis (Birks and Madsen, 1979). The pollen and macrofossil record suggests that after a phase with species-rich grasslands, tall-herb, fern, and willow communities, and some open vegetation present soon after the late-glacial, the vegetation cover became complete about 9000 B.P. The major vegetation types were species-rich grasslands, with tall-herbs and fern-dominated stands and small areas of birch and hazel scrub in sheltered areas, perhaps similar to the vegetation today surviving on ungrazed ledges and islands and in protected ravines. Soils were probably moderately fertile brown-earths. However, by 7700 B.P. *Calluna* heaths expanded, presumably as a result of soil acidification and podsolisation. The dominant vegetation was still grassland and tall-herb communities with *Angelica sylvestris*, *Filipendula ulmaria*, *Succisa pratensis*, and ferns including *Osmunda regalis*. There was never any extensive woodland around the site, only scattered birch, willow, hazel, and rowan scrub.

By about 5000 B.P. the widespread regional vegetation had become a mosaic of acid grassland, *Calluna* heath, and bog. Tall-herb and fern-rich communities became increasingly rare, presumably confined to areas of locally fertile soils. There is evidence for significant human influence on the vegetation from about 4000 B.P., leading to further expansion of heather moor, reduction of willow scrub and tall-herb stands, and spread of grassland and pasture with *Plantago lanceolata*. There is evidence for some cereal cultivation between 1700–1100 B.P., the time of colonisation of the Hebrides by Scots and Picts.

The above interpretation of the former extent of trees on Lewis has been challenged by Wilkins (1984) on the grounds that birch, pine, and willow wood-remains occur in peats near Little Loch Roag and these indicate 'the extensive growth of *Pinus* on Lewis prior to 4500 B.P.' (Wilkins, 1984, p. 258). Remains of birch, hazel, and, more rarely, alder and pine have been recorded from Lewis, Harris, Benbecula, and the Uists (see references in Birks and Madsen (1979) as well as Wilkins' (1984) own records), although the status of the pine records from the Uists is unclear (K. D. Bennett personal communication). There is, however, *no* 'disparity' (Wilkins, 1984) between the wood-remains on Lewis and our interpretation of the Roag pollen data. Both can indicate small areas of scrub in *local*, sheltered situations and a predominantly treeless *regional* vegetation. It is perfectly feasible to have sparse tree populations at densities of 0.25 trees ha$^{-1}$ (= 200 trees or less within 2–5 km radius of a pollen site) that are largely undetected pollen analytically (see Bennett (1985) for the basis of these calculations). The so-called 'disparity' is simply that pollen primarily provides an integrated record of *regional* vegetation over a large area, whereas macrofossils reflect strictly *local* patterns.

Pine stumps on Lewis date from 4800 – 3900 B.P., suggesting that pine was a late arrival there. Interestingly, pine reached Skye at about the same time (Birks and Williams, 1983). It became extinct over large areas of North-West Scotland, including Lewis and Skye, between 4300 – 3900 B.P. (see Birks, 1977; Bennett, 1984). Reasons for this widespread and spectacular demise are not fully understood, but a combination of climatic change and human activity may have accelerated the replacement of pine on flat and gently sloping ground by treeless blanket-bog. There is independent evidence for a change to a more oceanic climate with increased precipitation and strong winds at 4300 – 4000 B.P. (see Birks and Williams, 1983), possibly resulting from shifts in the Atlantic storm tracks due to changing positions and strengths of the Azores high and Iceland low. Such a climatic change would severely limit pine growth by causing waterlogging, encouraging bog expansion, and inhibiting regeneration by reducing the number of good seed-years. Interestingly, Ritchie (1985) proposes that massive sand-blowing and renewed machair formation on Pabbay began at about 4300 B.P. Machair development had been occurring earlier as a result of rising sea-level carrying glacially derived sand across the shallow, gentle off-shore shelf to create early machair beaches and dunes (Ritchie, 1979; Whittington and Ritchie, 1988). However, development was enhanced after about 4500 B.P. There is also evidence for increased storms and winds at this time elsewhere in the Outer (Simpson, 1966) and Inner (Birks, 1987) Hebrides and from Orkney (e.g. Keatinge and Dickson, 1979). Ritchie (1985) and Whittington and Ritchie (1988) argue that extensive machair has developed in the last 4000 years and that such a fundamental change in the geography and soils of the west coast would have had great influence on settlement history and patterns in this extreme marginal area. There is clearly a complex but poorly understood interaction between climatic change, vegetation dynamics, machair formation, sea-level rise, and human settlement in the Outer Hebrides that would repay further study (Whittington and Ritchie, 1988).

The former occurrence of coastal stands of birch, as indicted by submerged wood-remains of mid-post-glacial age off Pabbay, between the Sound of Harris and N. Uist, and on the west coast of Benbecula and S. Uist (Ritchie, 1966, 1979, 1985; Currie, 1979) also suggest that the frequency of westerly storms must have been less prior to 4000 B.P. As Currie (1979 p. 227) suggests, 'the stumps remaining are not likely to be an indication of forest, but rather the evidence of such sheltered locations where small woods survived the climatic conditions, often in areas which are now submerged by the sea'. Sea-level was 3–5 m lower between 8000 and 5100 B.P., as evidenced by abundant submerged terrestrial deposits along the west coast of the Uists (Ritchie, 1966, 1979, 1985).

Work currently in progress by K. D. Bennett and J. A. Fossitt is providing new and exciting insights into the vegetational history of the Uists. Pollen analysis from Loch Lang, S.Uist (Bennett and Fossitt unpublished) near the famous Allt Volagir ravine (Spence, 1960) show maximum tree pollen values of between 50% and 70% total pollen for the early and mid-post-glacial. The pollen frequencies suggest that woodlands of birch, hazel, oak, and alder were locally abundant with some elm, ash, and willow. Associates include *Ilex aquifolium*, *Hedera helix*, *Pteridium aquilinum*, and other ferns. This reconstruction contrasts with the inferred vegetational history at Little Loch Roag further north on the western side of Lewis. A similar contrast exists today in the cryptogamic flora of Lewis and the Uists, with several warmth-demanding species of shaded rocks and ravines (e.g. *Dryopteris aemula*, *Marchesinia mackaii*, *Lophocolea fragrans*, *Jubula hutchinsiae*) being present in the Uists but absent or very rare in Lewis. The Loch Lang pollen record and its clear indications of mixed birch-hazel-oak-alder woodlands in the early and mid-post-glacial suggest that these southern-atlantic cryptogams may indeed be relicts from the period of more extensive woodland cover in the mid-post-glacial and these shade-demanding species have survived locally in ravines, gullies, and sheltered block litters on the Uists.

Bog, grassland, and heathland expanded around Loch Lang in the later post-glacial to form the present predominantly treeless landscape of S. Uist. This expansion was almost certainly a result of a complex interaction between soil degradation, human impact, grazing, bog development, and, possibly, climatic change. When radiocarbon dates are available from Loch Lang and Bennett and Fossitt's investigations at other Outer Hebridean sites are completed, we will have a firmer factual basis for interpreting the vegetational history and the underlying factors that have influenced the temporal and spatial patterns of the past flora and vegetation in the Outer Hebrides.

Pollen spectra from shallow peat deposits associated with windblown sand on north-east Benbecula and southern Grimsay (Whittington and Ritchie, 1988) also indicate the possible local occurrence of oak, elm, and alder in mid-post-glacial times.

Local-scale studies in the Callanish area of Lewis (Bohnke, 1988) similarly indicate that small pockets of birch, hazel, willow, rowan, and aspen woodland with some *Lonicera periclymenum*, *Melampyrum*, and ferns occurred locally in sheltered areas. Bohncke proposes from detailed pollen analyses of several pollen profiles that human clearance of these pockets began as early as 7600, 5000, and 4200 B.P.. Almost total clearance occurred at about 3500 B.P. The Standing Stones of Callanish are thought to date from 3750 – 3500 B.P.

Walker's (1984) detailed diagram from Gleann Mor, St Kilda is unfortunately incomplete, with much of the mid-post-glacial missing. The pollen stratigraphy indicates that *Salix*-dominated vegetation, with a wide variety of herbs was prominent until about 6000 B.P. At about this time *Empetrum* and *Calluna* heath, grassland with abundant *Plantago lanceolata* and *P. maritima*, and fern-rich vegetation became widespread. Then there is a hiatus in the profile, with the period 5500 – 2000 B.P. missing, possibly as a result of drying out (Walker, 1983) or, more likely, as a result of prehistoric peat-cutting before 2000 B.P. (Birks, 1987; Whittington and Ritchie, 1988). From about 2000 B.P. the vegetation was maritime grassland and heath with abundant *Plantago maritima*, *P. lanceolata*, *Potentilla*, and *Ranunculus*. Walker interprets changes in the relative frequency of *P. maritima* pollen in terms of recent climatic changes, particularly changing storm-frequencies associated with the 'Little Ice Age' during the second millennium AD. In the absence of radiocarbon dates these interpretations must remain untested but attractive hypotheses. There is no evidence from pollen or macrofossils to indicate any tree or shrub growth on St Kilda except for *Salix*.

Although climatic changes undoubtedly occurred during the post-glacial, it appears that the principal determinants of regional vegetational change in the Outer Hebrides have been natural soil changes, with widespread leaching, podsolisation, and bog development beginning at about 7000 B.P. From about 5000 B.P. or even earlier (Bohnke, 1988), human influence has also been an important factor. In terms of climatic change, the major change in the Outer Hebrides appears to have been about 4000 B.P. leading to the extinction of *Pinus* in Lewis and Harris and the initiation or expansion of widespread machair development, probably as a result of increased storm frequency and westerly gales. Changes in wind appear to have been more important ecologically than changes in temperature or precipitation. It is likely that small areas of birch-hazel scrub were once more widespread in sheltered areas in Lewis prior to 4000 B.P., and that woodland of birch, oak, hazel, elm, and alder formerly occurred in S. Uist.

The report of a nut of *Trapa natans*, the water chestnut, from Loch Ceann a'Bhaigh, S. Uist (Heslop Harrison and Blackburn, 1946) would, if it is really of post-glacial age, indicate an enormous northward range-extension of *Trapa* (see Flenley *et al.*, 1975) and suggest a major climatic warming. The nut was not, however, found *in situ*, but washed up on the loch shore. Perhaps the nut is of interglacial age and redeposited from, as yet, undiscovered interglacial deposits within the catchment of the loch. In the absence of further work, this record remains enigmatic.

A total of 158 taxa have been identified as pollen, spores, or macrofossils from the Outer Hebrides. Of these, twelve (e.g. *Tilia*, *Fagus*, *Juglans*, *Abies*, *Picea*, *Ephedra*, *Sarcobatus vermiculatus*) are almost certainly the result of far-distance pollen-transport or secondary redeposition. Of the 146 taxa that, in terms of their present-day pollen representation, almost certainly grew on the Outer Hebrides, several are now extinct there. These include *Pinus sylvestris*, *Quercus*, *Ulmus glabra*, *Alnus glutinosa*, *Mercurialis perennis*, *Koenigia islandica*, *Helianthemum*, *Cannabis sativa/Humulus lupulus*, and *Anthoceros laevis*. The timing and reasons for these extinctions vary from taxon to taxon. Soil deterioration, competition, and climatic change may, as on Skye (Birks, 1973) have eliminated *Helianthemum* in the early post-glacial. Soil acidification, bog development, and scrub and woodland reduction in the mid- or late-post-glacial may have exterminated *Ulmus*, *Alnus*, *Quercus*, and *M. perennis*, whereas climatic change around 4000 B.P. probably caused the extinction of *P. sylvestris*. *K. islandica* could have been affected by competition and loss of habitat, as it is confined today to very open, mildly basic wet gravel flushes and, more rarely, fine-grained screes on Skye and Mull. Land-use changes may have eliminated *Cannabis sativa* (formerly cultivated on Lewis, – see Bohncke, 1988) and *Anthoceros laevis*.

Compared with fossil pollen spectra from Skye and elsewhere in the Inner Hebrides, spores of *Gymnocarpium dryopteris* are conspicuous in the Outer Hebrides by their absence. Although this plant occurs there today, its absence in the pollen record suggests that it was never common in the Outer Hebrides.

The vegetational and floristic history of the Outer Hebrides can be summarised as one of progressive impoverishment. Because of their insular ecology, area, distance from mainland sources, and limited habitat range the flora has always been restricted, simply because of the vagaries of dispersal, establishment, and extinction. The acidic bedrock and associated soils and extensive podsolisation and bog development, in conjunction with increasing storm frequency and human impact since about 4000 B.P. have lead to progressive impoverishment of the flora and vegetation over thousands of years resulting in the poor botanical diversity of the Outer Hebrides today.

*Acknowledgements*

I am particularly indebted to Keith Bennett and Julie Fossitt for allowing me to quote their unpublished results from Loch Lang; to Hilary Birks and Keith Bennett for reading the manuscript; and to Hilary Birks for word-processing.

# The Vegetation of the Outer Hebrides

R.J. PANKHURST

## Introduction

Several descriptions of vegetation of parts of the Outer Hebrides have already been written, and Currie (1979) provides a review and a commentary on these. During the later stages of the field surveys which were made for this Flora, the National Vegetation Classification project (NVC) was being carried out. The NVC is intended to provide a national standard for describing vegetation, in much the same way as a national Flora describes the species, and so it is clearly desirable to refer to the NVC in local Floras. At the time of writing, the NVC existed in draft form only, but we have been privileged to see this draft in order to prepare the following account. The following volumes were consulted:-

Saltmarsh
Maritime cliff communities
Calcicolous grasslands
Acid grasslands and uplands
Woodlands and scrubs
Heaths
Swamps and tall-herb fens
Mires and bogs
The following were not available:-
Aquatics
Weeds
Sand dunes and shingle

The resources of the NVC were limited, and they were not able to survey every vegetation type in the Outer Hebrides, so it is quite possible that there are more communities present than have been recorded. Our own recording efforts were concentrated on species, and we did not usually attempt to record vegetation. Nevertheless, it is worthwhile attempting to relate the NVC to what is known about the vegetation.

Existing descriptions of vegetation concentrate markedly on islands, rather than the mainland, in much the same manner as floristic exploration has done. The most general account of vegetation so far published is that of Wilmott (1945) for Uig, but that does not of course use the modern methods for vegetation description. Useful data with descriptions and records of communities from the machair have been provided by Imogen Crawford (pers. comm.). Apart from the NVC, useful modern descriptions of Scottish vegetation types are to be found in McVean and Ratcliffe (1962), Birks (1973) and Birse and Robertson (1976). Unfortunately, none of these has any data for the Outer Hebrides. Papers which include information on the vegetation of particular islands are given in an appendix.

## Definition of communities

The definition of a **community** does not include any features of the habitat; it depends on the frequency and abundance of species only. A community is roughly the equivalent of a Braun-Blanquet Association, as used by the Montpellier school of phytosociologists. For convenience however, the following account is organised in terms of habitats, with definitions and lists of the communities known or suspected to occur within them. See Robertson (1984) for a handy key to vegetation types in Scotland, which includes some of the communities of the Outer Hebrides.

A community is defined in terms of its **constant** species i.e. the species which occur in 61–100% of quadrats. This means that an individual quadrat may not necessarily contain all of the constants, although they are likely to occur nearby. Communities may sometimes be divided into **sub-communities** by **differential** species. A differential species is one which is strictly confined to a particular sub-community, and not found in others. **Associate** species are those which are not constant but which occur throughout the community with a frequency of 60% or less, and which do not favour any particular sub-community. Rare species do not form part of the definition of a community, although it may be useful to state in which community the rarity occurs in order to fix its ecological preference. The Braun-Blanquet Associations are defined by the obligatory presence of rare species, the so-called 'indicator species'. This notion is impractical in Britain because the vegetation is often considerably poorer in species than on the continent, owing to the more northerly latitude, so that there are insufficient 'indicators' upon which reliance may be placed. A striking example of this is the Cakile maritima – Honkenya peploides community which is otherwise known as the Mertensio-Atriplicetum laciniatae association, from which Mertensia is nearly always absent in Scotland!

In the following account, localities for vegetation types have been given where these are present on the dot maps of the NVC. Otherwise, no attempt has been made to equate the vegetation described in the older publications with their modern equivalents.

### Saltmarsh

A valuable reference for saltmarsh vegetation is Adam (1981). See also C. H. Gimingham in Burnett (1964). Pitkin et al. (1983) describe a number of brackish communities in South Uist, but these are not clearly comparable with the NVC. Law and Gilbert (1986) have made a survey of saltmarsh in the Western Isles, and their community definitions appear to be closely similar to those of the NVC.

Zostera communities
Stands of *Zostera marina* or *Z. angustifolia*. Defined by the presence of a single species of higher plant.
Vatersay to Sound of Harris    14–16,18,19,21,22

Ruppia communities
*Ruppia maritima* or *R. cirrhosa* occurs in the Outer Hebrides as a species of brackish lochs, and perhaps not as a constituent of saltmarsh as referred to by Adams(1981) and in the NVC. Nevertheless, the NVC does record this community in the Outer Hebrides (South Uist to north Lewis).

Spartina townsendii community
Defined by presence of *Spartina townsendii* or *S. anglica*. Occurs towards the seaward edge of saltmarshes. Luskentyre, S.Harris.    12

Sarcocornia (Salicornia) perennis community
This species is confirmed from only one locality in the Outer Hebrides, although plants identified just to the genus *Salicornia* alone are recorded quite widely. Nevertheless it is doubtful whether the vegetation in which it occurs is sufficiently well-developed that it could be identified with the community described in the NVC.

Puccinellia maritima community
Defined by *Puccinellia maritima* as a constant species. In the NVC, the *P. maritima* dominated sub-community is recorded from Lewis. Other species may occur at higher levels such as *Triglochin maritima*, *Spergularia media*, *Plantago maritima* and *Armeria maritima*. Other of the sub-communities defined by the NVC probably also occur in the Outer Hebrides, see Law and Gilbert (1986).

Juncus gerardi community
Defined by *Juncus gerardi* as a constant. Several sub-communities are recorded in the Outer Hebrides:
a)  *Festuca rubra – Glaux maritima* sub-community (Lewis, S.Harris, N.Uist). *J. gerardi*, *F. rubra* and *G. maritima* are constants along with *Plantago maritima* and *Agrostis stolonifera*, with frequent *Triglochin maritima* and *Armeria maritima*.
b)  *Leontodon autumnalis* sub-community (Lewis, N.Uist). *J. gerardi*, *Agrostis stolonifera*, *F. rubra*, *Trifolium repens* and *G. maritima* are the constants, with common *L.autumnalis* and *Carex distans*.
c)  *Carex flacca* sub-community (Lewis). *J. gerardi*, *Agrostis stolonifera*, *Festuca rubra*, *Plantago*

*maritima, Trifolium repens* and *Leontodon autumnalis* are the constants, with common *Glaux maritima*.

Juncus maritimus saltmarsh
    Defined by *Juncus maritimus* clumps with an understorey of *Agrostis stolonifera, Festuca rubra, Glaux maritima* and *Juncus gerardi*. Not recorded in the NVC, but present in the Uists according to Pitkin et al. (1983).

Blysmus rufus community
    Defined by constant and dominant *Blysmus rufus* with constant *Juncus gerardi, Glaux maritima* and *Agrostis stolonifera*, with frequent *Triglochin maritima*. Not specifically recorded in the NVC, but occurrences of this community can probably be equated to records for *B.rufus*. What is presumably this community is frequently referred to by Law and Gilbert (1986). To be expected on sandy or gravelly shores where there is flushing with fresh or brackish water.

Eleocharis uniglumis saltmarsh
    This community is very similar to the last, except that *Eleocharis uniglumis* replaces *Blysmus rufus*. This community is recorded for the Uists by Pitkin et al. (1983) and Law and Gilbert (1986).

**Fig. 9**   *Machair and dunes, Halaman Bay, Barra*
*3 June 1976 (A. Currie)*

*Shingle strandline and sand-dune*

There is an account of sand-dunes at Luskentyre in S.Harris by Gimingham et al. (1948). Pitkin et al. (1983) describe a series of dune and dune-slack communities for South Uist, with detailed distribution maps.

Cakile maritima- Honkenya peploides community (S. Harris, Benbecula, Barra) on sandy foreshores. Defined by the two named species, with *Elymus farctus* and *Atriplex* species frequently present.

Elymus (Agropyron) foredune community (N.Uist, S.Uist, Barra). The constant is *Elymus farctus (Agropyron junceiforme)* while *Ammophila arenaria* is more or less absent.

Ammophila yellow dune, with abundant or dominant *Ammophila arenaria* (Barra to Lewis). Ammophila may occur on its own or with a variety of other species such as *Festuca rubra* and *Senecio jacobea* (Huiskes, 1979). The latter may be what Robertson (1984) calls northern marram grass dune pasture.

Festuca rubra – Galium verum dune (Barra to Lewis). *Festuca ovina* is absent from this community, *Koeleria gracilis* and *Euphrasia* species are frequent. This may be the same as the eyebright-red-fescue dune of Robertson (1984).

Potentilla anserina – Carex nigra community (Lewis, S.Harris, Uists, Monach Is.). *Carex nigra* is abundant or dominant and *C. flacca* and *Potentilla anserina* abundant or frequent. This might be regarded either as a kind of dune slack or as a type of poor fen.

Also found on dunes is the Variegated horse-tail community, which may occur where there are records of *Equisetum variegatum*, although it is not actually recorded from the Outer Hebrides. The other constant species is *Anagallis tenella* and the community is generally rich in species.

The most important and famous of all Hebridean habitats, the machair, falls under the sand-dune section. Machair is Gaelic for 'plain', and Ranwell (1974) defines it as 'a type of dune pasture (often calcareous) subject to local cultivation developed in wet and windy conditions in the north and north west of Scotland.' There are many general publications about machair, but until the appropriate volume of the NVC appears, there is not much precise information about machair vegetation available. Randall (1976) mentions machair communities, which he calls either;
a)  Festuca rubra community with *Bellis perennis, Plantago lanceolata* and *Achillea millefolium*, or
b)  mature machair, Festuca rubra community but with high proportion of *Trifolium pratense*.
    According to Crawford (pers.comm.), natural machair is a Festuca rubra – Galium verum grassland with characteristic *Ranunculus acris, Euphrasia officinalis, Bellis perennis* and *Plantago lanceolata*. The lower and damper machair has increasing amounts of *Euphrasia, Carex flacca, Potentilla anserina* and *Gentianella campestris*, and is transitional to salt-marsh. Drier hummocks support *Carex arenaria, Sedum acre* and *Senecio jacobaea*. Hollows contain dense *Festuca rubra, Agrostis stolonifera* and *Potentilla anserina*, which is one of the communities of mesotrophic grassland. Inland the machair grades into mesotrophic grassland and then to fen and swamp communities.

*Maritime cliff*

Armeria maritima – Ligusticum scoticum maritime rock crevice community. *A. maritima* and *L.scoticum* are constant, as also is *Festuca rubra. Plantago maritima, Sedum rosea* and *Silene maritima* are frequent. *Asplenium marinum* is occasional but may be plentiful if sheltered. The same community, under the name of the Armeria maritima – Ligusticum scoticum association, is recorded for two localities in Lewis by Malloch and Okusanya (1979). (N.Harris, Lewis)

Sedum rosea – Armeria maritima maritime cliff ledge community. The constant species are *S. rosea, A.maritima* together with *Festuca rubra* and *Rumex acetosa. Plantago maritima* and *P. lanceolata* with *Holcus lanatus* are frequent, and since this community is ungrazed, it creates a luxurious herbaceous vegetation. *Angelica sylvestris* and *Silene dioica* may be evident. (N.Harris, Lewis).

Festuca rubra – Armeria maritima maritime grassland. *F. rubra* and *A. maritima* are the constants. Other frequent species are *Agrostis stolonifera, Plantago maritima, Daucus carota* and *Silene maritima*. (N.Uist to Lewis). Two sub-communities may occur in the Outer Hebrides:
a)  Ligusticum scoticum sub-community. *L. scoticum* is the additional constant, and there may be also *Agrostis stolonifera* and *Holcus lanatus*.

b) Plantago coronopus sub-community. The additional constants are *Agrostis stolonifera, Plantago coronopus* and *P.maritima*

Festuca rubra – Holcus lanatus maritime grassland. Characteristic of somewhat sheltered situations, such as towards the clifftops or on lee slopes. The constant species are *F. rubra* and *H. lanatus* with *Plantago lanceolata* and *Armeria maritima*. *Plantago maritima, Rumex acetosa* and *Trifolium repens* are frequent. (Lewis). Sub-communities known from Scotland are:
a) Plantago maritima sub-community. The additional constants are *Lotus corniculatus, Plantago maritima* and *Trifolium repens*. *Scilla verna* is frequent here. Robertson (1984) calls this the vernal squill maritime pasture.
b) Anthoxanthum sub-community. *Armeria maritima* is now no longer constant, though frequent, and the additional constants are *Anthoxanthum odoratum, Agrostis capillaris, Rumex acetosa, Poa subcaerulea* and *Potentilla erecta*.

Festuca rubra – Plantago maritime grassland. The constants are *F.rubra* with *P.coronopus, P. lanceolata, P. maritima* and *Agrostis stolonifera* (N.Uist to Lewis). There are three sub-communities:
a) Armeria maritima sub-community. *A. maritima* is the only additional constant.
b) Carex panicea sub-community. The additional constants are *C. panicea, Lotus corniculatus, Leontodon autumnalis* and *Euphrasia* spp.
c) Schoenus nigricans sub-community. (Harris and Lewis). The constants are *S. nigricans* (as the prostrate marine ecotype), *Carex panicea, C. serotina, Molinia caerulea, Danthonia decumbens, Euphrasia* spp. and *Potentilla erecta*.

*Mires and bogs*

A **mire** is defined here in the sense of Ratcliffe in Burnett (1964) as a habitat in which the water-table is at or near the surface but where there is lateral water movement. It is distinguished

**Fig. 10** *Moorland at Barvas, Lewis*
1 June 1976 (A. Currie)

from a **bog** where the water arrives solely in the form of rain from above, and there is no run-off. For accounts of mires in Lewis, see Goode and Lindsay (1979), Hulme and Blyth (1984) and Hulme (1985).

Sphagnum auriculatum bog pools (Lewis, north and south Harris). The constant species are *Eriophorum angustifolium, Menyanthes trifoliata, Sphagnum auriculatum* and *S. cuspidatum*, while *S. recurvum* is rare. *Carex limosa* and *Rhynchospora alba* are often found in this community.

Hulme and Blyth (1984) recognise a Sphagnum cuspidatum – Utricularia minor vegetation type for recolonised bog pools in Lewis.

Eriophorum angustifolium bog pools (Lewis, S. Harris). The *Eriophorum* is frequent or abundant in bog pools dominated by *Sphagnum* species in which other vascular plant species only occur occasionally.

Carex echinata – Sphagnum recurvum/auriculatum mire. Defined by constant *Carex echinata*, the *Sphagnum* species and the moss *Polytrichum commune*. The Carex echinata sub-community (Lewis) is dominated by *C. echinata* with *C. nigra* and *C. panicea* being less common, and the grasses *Molinia caerulea* and *Agrostis canina. Eriophorum angustifolium* is frequent in this sub-community.

Scirpus cespitosus – Erica tetralix wet heath (Lewis, north and south Harris). This is a rather variable type of vegetation and is characterised by constant *Molinia caerulea, Scirpus cespitosus, Erica tetralix, Potentilla erecta* and *Calluna vulgaris*, but any one or even two of these species might be missing. *Polygala serpyllifolia, Narthecium* and *Eriophorum angustifolium* are preferential in this community.
a)   Carex panicea sub-community. This typically has much more *Molinia* and *E.tetralix*, especially the former, while the other two species are relatively scarce. A range of different *Carex* species and *Drosera rotundifolia* are preferential, and *Schoenus nigricans* and *Myrica gale* are locally abundant.
b)   typical sub-community. This sub-community is highly variable in respect of the dominant species but the *Carex* or *Cladonia* species are absent or uncommon. Again *Myrica gale* may occur.
c)   Cladonia sub-community. All the dominants have high frequency here with *Calluna* the most dominant, and *Cladonia* species are frequent.
Hulme and Blyth (1984) recognised several vegetation types which might be forms of the above communities involving *Calluna vulgaris* and *Potentilla erecta* in disturbed peatland in Lewis.

Scirpus cespitosus – Eriophorum vaginatum blanket mire (Lewis and Harris). The constant species are *Calluna vulgaris, Erica tetralix, Molinia caerulea, Scirpus cespitosus* and *Eriophorum vaginatum.* The high frequency of *E. vaginatum* is one of the distinctions between this community and the Scirpus-Erica wet heath, where *E. angustifolium* is characteristic.
a)   Drosera rotundifolia – Sphagnum sub-community. This is dominated by either *Calluna* and *Scirpus* or *Calluna* and *Molinia. Drosera rotundifolia* and *Myrica* are preferential here, although the latter is often absent. The most abundant Sphagna are *S.capillifolium* and *S. papillosum*.
b)   Cladonia sub-community. This is usually dominated by *Calluna* and *Scirpus. Drosera rotundifolia* is uncommon and Sphagna are mush less abundant, while *Cladonia* species and *Racomitrium* are much more obvious.

Iris pseudacorus – Filipendula mire. This community is not specifically recorded from the Outer Hebrides, but Adam (1976) states that it is probably ubiquitous on marshes in western Scotland. *Iris pseudacorus* is constant with frequent *Filipendula ulmaria* and there are usually other weedy taxa present such as *Galium aparine, Poa trivialis* and *Urtica dioica*.

*Heaths*

For a general discussion of heaths in Scotland, see Gimingham in Burnett (1964). The distinction between a mire or bog association containing ericaceous plants and a heathland proper may not be altogether obvious, but the heath vegetation described under this heading tends to occur in drier habitats.

Calluna vulgaris – Scilla verna heath. This community occurs on maritime cliffs within reach of salt spray. The constant species are:

*Calluna vulgaris*
*Erica cinerea/tetralix*
*Festuca ovina*
*Holcus lanatus*
*Hypochaeris radicata*
*Lotus corniculatus*
*Plantago lanceolata*
*P. maritima*
*Potentilla erecta*
*Scilla verna*
*Thymus praecox*

*Calluna vulgaris* is the most frequent shrub, accompanied by *E. cinerea* on drier soils but in north Britain *E. tetralix* or even *Empetrum* often replace it. Although the ericaceous shrubs are a consistent part of the community, they are usually much reduced in stature by exposure and grazing, and may be only a few centimetres high.

a)   Erica tetralix sub-community (Lewis, N. Harris and N. Uist). The dominant shrubs are *Calluna* and *E. tetralix*. The commonest grass is *Festuca rubra* and the commonest sedge is *C. nigra*.

b)   Calluna vulgaris sub-community (Lewis). This is more impoverished than other sub-communities, and is dominated by *Calluna* accompanied by *E. cinerea*. *Festuca ovina* is more frequent than *F. rubra*.

## Woodlands and scrub

For a general reference on Scottish woodland vegetation, see McVean in Burnett (1964). There is little in the Outer Hebrides at the present time which can be described as natural woodland. There are a number of plantations of commercial forestry (Blake, 1966) and some parks, as at Lews Castle in Stornoway. There are a number of restricted areas of scrub, either on small islands in lochs or in river ravines, see Spence (1979), and Currie (1981b).

There is no record of wood or scrub communities in the Outer Hebrides in the National Vegetation Classification, but the Dryopteris dilatata – Rubus fruticosus sub-community of the Betula pubescens – Molinia caerulea woodland community closely resembles the birch wood in South Uist described by Bennett and Fossitt (1988). The scrub community of Allt Volagir, also in South Uist and mentioned in the same paper, is somewhat similar, but *Betula pubescens* is absent. The constant species of small islands in lochs of Lewis, Harris and the Uists are *Sorbus aucuparia, Salix aurita* or *S. cinerea, Rubus fruticosus* and *Dryopteris dilatata* and these might also represent a degraded form of the same sub-community.

Spence (1979) also describes what he calls the Corylus – Primula – Ranunculus ficaria community from rock ledges in Allt Volagir, S. Uist. This does not seem to correspond to any community published elsewhere, but of course the environment is rather marginal for woodland growth at this locality. Other characteristic species present apart from the three by which the community is named are *Populus tremula, Betula pubescens, Sorbus aucuparia, Salix aurita/cinerea* and *Epilobium montanum*.

## Mesotrophic grassland

There are no records of mesotrophic grassland for the Outer Hebrides in the NVC, but it is known that the following communities occur:

The Cynosurus cristatus – Centaurea nigra meadow, which is also known from the seaboard of western Scotland.

Arrhenatherum elatius coarse grassland.

Festuca rubra – Agrostis stolonifera – Potentilla anserina inundation grassland (see above under machair).

Cynosurus cristatus – Caltha palustris flood pasture.

There is also the Lolium perenne – Cynosurus cristatus pasture, in which *Trifolium repens* is usually abundant, with frequent *Holcus lanatus*, *Festuca rubra* and *Poa pratensis*. This community is cultivated land and grasses will probably have been sown. It might be identified on or near machair.

### Weeds

The NVC for weeds was not available at the time of writing, but two named communities which are likely to be recognised are:

Pineappleweed trample community (Polygonum aviculare – Matricaria matricarioides), on compacted soil in farmyards, gateways and along tracks. The constants are *Matricaria matricarioides*, *Polygonum aviculare* and *Poa annua*.

Great plantain trample community (Lolium – Plantago major) in similar places to the last. *Plantago major* is constant and *Poa annua*, *Trifolium repens*, *Lolium perenne* and *Cynosurus cristatus* are frequent.

Silverside (1977) describes the Chenopodium album – Viola tricolor subsp. curtisii association from the cultivated machair of North and South Uist. The other diagnostic species are *Avena strigosa*, *Anchusa arvensis*, *Erodium cicutarium*, *Galium verum*, *Thalictrum minus*, *Tripleurospermum maritimum* and *Honkenya peploides*. This community is probably the result of cereal cultivation. Another distinctive weed community occurs after the cultivation of potatoes, containing *Myosotis arvensis*, *Sinapis arvensis*, *Stellaria media*, *Sonchus asper* and *Papaver rhoeas*, with a smaller component of *Viola tricolor*. If left unploughed, these communities are eventually replaced by a Festuca rubra – Trifolium repens sward.

**Fig. 11** *North Uist seen from North Lee*
*13 June 1978 (A. Currie)*

*Aquatics*

The NVC for aquatic plants was not available at the time of writing, but see Spence in Burnett (1964). The surveys carried out by the Royal Botanic Garden, Edinburgh (1983, 1984 and 1985), contain a great deal of information about the plants in lochs from South Uist to Lewis. Spence et al. (1979) record the presence of the Potamogeton filiformis – Chara aspera association in Loch Grogarry, S. Uist. This is also recorded by the RBG, Edinburgh survey of 1983 for Loch Aird an Sgairbh, South Uist. Other species associated with the dominants are *Myriophyllum alterniflorum* and *Littorella uniflora*.

Other named communities which are likely to be recognised are:

Water Lobelia community (Isoetes – Lobelia dortmanna), at the waters edge or submerged in sandy or peaty lochs and lochans. Open stands of *Lobelia dortmanna* with *Eleocharis multicaulis* and *Ranunculus flammula* (presumably the same as the community reported in RBG, Edinburgh survey for 1984).

Marsh St. John's wort community (Hypericum elodes – Potamogeton polygonifolius) in peaty pools. Constants are *Hypericum elodes* and *Potamogeton polygonifolius*.

Amphibious bistort community (Polygonum amphibium community) in mesotrophic and eutrophic low-lying lochs and lochans. Constant *Polygonum amphibium* with various species of *Potamogeton*.

White water-lily community (Nymphaea alba) in more oligotrophic lochs and lochans, dominated by the floating leaves of *Nymphaea alba*.

Ivy-leaved Water crowfoot community (Ranunculus hederaceus). Constant *Ranunculus hederaceus* and frequent *Callitriche stagnalis* and *Juncus bufonius*.

*Swamps and tall-herb fens*

None of these communities is specifically recorded in the NVC for the Outer Hebrides (but see Spence in Burnett, 1964, and the surveys by the Royal Botanic Garden, Edinburgh, 1983, 1984, and 1985). The species involved are characteristically tall perennial monocots which tend to form pure stands on the edge of water bodies. There are a number of species which occur in this way and hence define communities in a rather trivial sense, since there are few or no associates. Communities based on pure stands of *Equisetum fluviatile, Cladium mariscus, Schoenoplectus lacustris, Scirpus maritimus, Eleocharis palustris, Phragmites australis, Phalaris arundinacea* and *Glyceria fluitans* probably all occur in the Outer Hebrides. The Carex rostrata swamp has a sub-community composed of pure *Carex rostrata*.

Pitkin et al. (1983) and Chamberlain et al. in the 1984 survey by the Royal Botanic Garden, Edinburgh, both record the presence of both poor fen and rich fen in South Uist and Lewis. Poor fen is defined in the NVC as a Carex rostrata – Potentilla palustris community. The constant species are *Carex rostrata, Potentilla palustris, Galium palustre*, and *Menyanthes trifoliata*. Rich fen is the Carex rostrata – Equisetum fluviatile sub-community of this. Both poor and rich fen occur in the form of 'schwingmoor' i.e. as a floating mat of vegetation over water. Hulme and Blyth (1984) recognise a kind of poor fen from Harris and Lewis which is dominated by *Menyanthes* but also contains *Carex limosa* and *Juncus bulbosus*.

Both the above sources also refer to what is evidently the Carex paniculata sedge swamp of the NVC, found in Benbecula and South Uist.

*Acid grasslands and uplands*

For general reference, see King and Nicholson in Burnett (1964) and Birse and Robertson (1976). The climate of the Outer Hebrides is such that much of the moorland vegetation belongs to the one of the mire types, and grades into maritime heath near the sea. Acid grasslands on dry or well-drained soils are relatively scarce. Most of the following communities are not actually recorded for the Outer Hebrides in the NVC (unless explicitly stated) but might well occur.

Festuca ovina – Agrostis capillaris – Galium saxatile grassland (siliceous grassland). The constant species are *Agrostis capillaris, Anthoxanthum odoratum, Festuca ovina, Galium saxatile* and *Potentilla erecta*. *Festuca vivipara* is not present, *F. rubra* may occur, and *Nardus, Deschampsia flexuosa* and *Molinia* only occur in small quantity. This community occurs in pastures on

**Fig. 12**   *Sròn Scourst and Oreval, North Harris*
*6 September 1978 (A. Currie)*

better-drained and base-poor soils in north-west Britain and mostly at low altitudes. Apart from the typical sub-community, there are:
a)   the Holcus lanatus – Trifolium repens sub-community. According to Hulme and Blyth (1984), reclaimed peatland in Harris and Lewis is characterised by pastures with *Trifolium repens, Holcus lanatus, Agrostis canina* and *Anthoxanthum*, which comes close to this sub-community. These pastures are created by adding shell sand and fertilisers and reseeding with grass mixtures and *Trifolium*.
b)   Luzula multiflora – Rhytidiadelphus loreus sub-community. Other species which commonly occur are *Agrostis canina, Deschampsia cespitosa* and *Nardus*.
c)   Vaccinium myrtillus – Deschampsia cespitosa sub-community. *Festuca ovina, Nardus* and *Deschampsia cespitosa* are common here.

Nardus stricta – Galium saxatile grassland. Constant species are *Agrostis capillaris, Festuca ovina/vivipara, Galium saxatile, Nardus stricta* and *Potentilla erecta*, but the *Nardus* is the most dominating species. Sub-communities include:
a)   species-poor sub-community, with *Nardus* as the most abundant plant, with frequent *Deschampsia cespitosa* and *Festuca ovina*.
b)   Agrostis canina – Polytrichum commune sub-community, in which there is somewhat less *Nardus*.
c)   the Carex panicea – Viola riviniana sub-community, which is distinctly richer in species.
d)   Calluna vulgaris – Danthonia decumbens sub-community, in which *Nardus* dominates, and where *Festuca ovina* and *Agrostis capillaris* are often abundant.
e)   and the Racomitrium lanuginosum sub-community (North Harris and Lewis), where *Nardus* still dominates along with much *Racomitrium* but few other grasses.

Juncus squarrosus – Festuca ovina grassland. These are also the constant species, together with the moss *Polytrichum commune*. The community typically occurs on peaty soils on gentle slopes or plateaus at higher altitudes. The sub-communities are:
a)  Sphagnum sub-community, also known as the Juncus squarrosus bog.
b)  Carex nigra – Calypogeia trichomanis sub-community
c)  Vaccinium myrtillus sub-community
d)  Agrostis capillaris – Luzula multiflora sub-community.

Carex bigelowii – Racomitrium lanuginosum moss-heath. The constant species are *Carex bigelowii, Deschampsia flexuosa, Festuca ovina/vivipara, Vaccinium myrtillus, Racomitrium* and the lichen *Cladonia uncialis*. Other species which may occur are *Alchemilla alpina* and *Salix herbacea*. This community is characteristic of windswept and could-ridden plateaus and hilltops in cold and humid north-west Britain. Although not recorded in the NVC, it occurs on the summits of hills in Harris and the Uists. Sub-communities include:
a)  the typical sub-community, and
b)  the Galium saxatile sub-community.

Luzula sylvatica – Vaccinium myrtillus tall-herb community. The constants are *Deschampsia flexuosa, Galium saxatile, Luzula sylvatica* and *Vaccinium myrtillus*. Quite a range of other species may occur, especially ferns. The community is confined to inaccessible cliffs or sides of ravines where grazing and burning are unlikely, but where trees cannot grow. This community has obvious affinities to the scrub vegetation of small islands and ravines described above, but is not recorded for the Outer Hebrides in the NVC. The sub-communities are:
a)  the Dryopteris dilatata – Dicranum majus sub-community
b)  the Anthoxanthum odoratum – Festuca ovina sub-community, and
c)  species-poor sub-community.

Luzula sylvatica – Geum rivale tall-herb community. The constant species are *Angelica sylvestris, Deschampsia cespitosa, Geum rivale, Luzula sylvatica* and *Sedum rosea*. Like the last community, it is confined to cliffs where it is protected from grazing and burning. The Primula vulgaris – Hypericum pulchrum sub-community is recorded in the NVC from North and South Uist. It often contains *Calluna* and a variety of ferns, especially *Blechnum*.

Oreopteris limbosperma – Blechnum spicant community. The constants are *Blechnum spicant, Galium saxatile, Oxalis acetosella, Potentilla erecta* and *Thelypteris limbosperma*. The community is characteristic of peaty soils on steep sheltered slopes at low to moderate altitudes in north-west Britain, and so is likely to occur in the Outer Hebrides.

Pteridium aquilinum – Galium saxatile community. The constants are *Festuca ovina, Galium saxatile, Potentilla erecta* and *Pteridium*. This is the formal description of the familiar bracken patch of heath or moorland. The community is not recorded in the NVC for the Outer Hebrides, but is very probably present where bracken occurs.

# The History of Botany in the Outer Hebrides

ANDREW CURRIE

*Nature Conservancy Council, Broadford, Skye*

## Introduction

The history of botanical exploration in the Outer Hebrides may conveniently be divided into four periods. First, the period until 1830 was the time of the early travellers; second, from 1830 until 1930 was the period of early botanists and naturalists; third, 1930 until 1980 was the period of the first ecological studies and also of Professor J. W. Heslop Harrison and Miss M. S. Campbell; and finally the 1980s, the period of the Integrated Development Programme (1982–1987), which gave rise to more intensive research. A valuable starting point for a study of the flora of the Outer Hebrides is the paper by Dr. H. R. Fletcher (1959) entitled 'Exploration of the Scottish Flora', in which he discusses plant recording in that group of islands. During the following year, N. Douglas Simpson (1960) printed privately 'A Bibliographical Index of the British Flora', including a list of published sources for the Hebrides; V.C. 110. A brief outline of the history is given in my own paper (Currie, 1979).

## 1. Period until 1830

Early references to the flora are present incidentally or by inference in accounts by early travellers to the islands or other writers about the islands. The earliest of these was Sir Donald Monro (1884), High Dean of the Isles, who wrote in 1549 (Darling & Boyd, 1964). Other early writers include Smollett (1768), Heron (1794), Walker (1812), Walford (1818) and MacCulloch (1819 & 1824), but there were several more, including those listed by Cooper (1979). Little other than isolated records can be established from most of these, although there are some fine descriptive passages. Simpson (op. cit.) refers to the tour by Pennant (1774 & 1776), but Slack (1986) makes it clear that Thomas Pennant and his companion the Rev. John Lightfoot did not visit the Outer Hebrides. It would appear therefore that the first traveller to offer substantial plant records was Martin Martin (1698 & 1703).

Martin Martin was a Skye man. Traditional Clan MacLeod lands included Harris and parts of Lewis, and at the time of his visit to St. Kilda in 1695, the MacLeods of Skye held the title to that isolated archipelago (Cooper, 1970 & op. cit.). Martin was for some years tutor to the MacLeods of Skye, and had also been at one time a scholar at Leiden University in the Netherlands. In 1695, he toured both the Inner and the Outer Hebrides, visiting many of the islands personally.

In 1698, Martin published 'A Late Voyage to St. Kilda, the Remotest of all the Hebrides', and five years later in 1703 there appeared the first edition of 'A Description of the Western Islands of Scotland'. These accounts contain many references to plants, and probably the earliest records of the species named. In his important paper, Perring (1953) extracted these records and listed them island by island. He had to translate early botanical, English, Scots and Gaelic names into their modern botanical equivalents, although it was not always possible to be precise. In the lists below, names used by Martin are followed by Perring's identification where this is specific. Plants other than phanerogams have been omitted.

From Lewis are listed Birch, Hazel (*Corylus avellana*), Nettle, Reed (probably *Phragmites australis*), Golden Rod (*Solidago virgaurea*) and All Heal (may be *Stachys arvensis*). From Harris there are Daisy (*Bellis perennis*), Clover, Wild White Carrot (*Daucus carota*) and Allium Latifolium (*Allium ursinum*). Milkwort is reported on Hermetray. In North Uist, there are Flamula Jovis (*Ranunculus flammula*), Shunnish; Suinas or Shunnas (*Ligusticum scoticum*) and Alexander (*Smyrnium olusatrum*). The latter was probably a garden escape, and is not found wild in V.C. 110 today. From Barra only Curran-Petris (*Daucus carota*) is recorded.

The St. Kilda list is as follows:– Sorrel (*Rumex acetosa*), Lapathum vulgare (*Rumex obtusifolius*), Scurvy Grass (*Cochlearia officinalis*), Mille-foil (*Achillea millefolium*), Bursa pastoris (*Capsella bursa-pastoris*), Silver weed or argentine (*Potentilla anserina*), Plantine, Sage (*Teucrium scorodonia?*),

Chicken weed (*Stellaria media*), All Hail or siderites (This may be All Heal (see earlier) or may be *Galeopsis tetrahit*), Sea Pink (*Armeria maritima*), and Tormentil (*Potentilla erecta*). Perring comments that all but *Teucrium scorodonia* have been recorded from St. Kilda this century.

The Rev. Alexander Buchan was minister on St. Kilda from 1705 until his death in 1730. He left behind him a manuscript which was subsequently published (Buchan, 1752). In this he refers to a St. Kilda man who had occasion to visit Harris and later Skye, and of that trip Buchan writes thus:– 'One of the things he and they with him wondered at most was, the Growth of Trees, they thought the Beauty of Leaves and Branches admirable, and how they grow to such a Height above Plants was far above their Conception. For there grow no Trees, no not so much as a Shrub on St. Kilda'.

Professor Mark Louden Anderson (1967) researched fully the early evidence of trees in the Outer Hebrides, and I quote freely from him. 'MacCulloch (1824) the geologist says that there were no trees on Uist'. And again, 'Of Lewis . . . they have long lost their claim to the name of the Birken Isles, as no birch now occurs there. On the other hand, Heron (1794) says that Lewis was destitute of wood except for a few birches and hazels'. MacCulloch (op. cit.) also mentions 'a few stunted plants of poplar, *Populus alba*', but these must surely refer to Aspen (*Populus tremula*). Beveridge (1911) refers to former evidence of trees based upon Gaelic place-names.

Many writers refer to tree remains in peat (Martin, (1703), MacCulloch (1819), MacGillivray (1830) and Niven (1902)). Once again I acknowledge Anderson (op. cit.) for the following:– 'Geikie (1894) states that in Lewis, fully-grown oak, alder, birch and especially Scotch fir, had been found in the bogs and MacCulloch (1824) in respect of the peat in North Uist, includes oak with pine, alder and birch amongst the trees most frequently found'. Anderson (op. cit. Appendix D) has assembled records from the New Statistical Account (1845) of occurrences of woody remains in peat bogs, including those from Inverness-shire (North Uist, South Uist, Barra) and Ross and Cromarty (Stornoway, Barvas, Lochs). He refers also to the island tradition that woods were destroyed by Viking invaders (Danes, Norwegians). Beveridge (1926) and Elton (1938) described submerged forests off Vallay, Pabbay and other islands of the Sound of Harris, while Ritchie (1966, 1979 & 1985) describes intertidal areas between the sound of Harris and South Uist where evidence of past woodland may be seen. Professor H. J. B. Birks deals fully with this topic and with floristic and vegetational history elsewhere in this volume.

## 2.    1830–1930

The date 1830 was chosen because it was during that year that the first true naturalist wrote. William MacGillivray was born in Old Aberdeen, but went to Harris at the age of three and spent eight childhood years there. He wandered widely throughout Scotland during his university days in Aberdeen and explored what he called 'the desolate isles of the west'. His subsequent account of the Outer Hebrides (MacGillivray, 1830) is one referred to by Campbell (1937) as outstanding. His eldest son, John, wrote about St. Kilda (MacGillivray, 1842) and William MacGillivray himself went on to become Professor of Natural History and Lecturer on Botany in Marischal College and University, Aberdeen.

I am indebted yet again to Anderson (op. cit.) for the following quotation. 'W. MacGillivray (1831) gives a very important account of the Outer Hebrides. He writes: "Speaking generally, one might pronounce these islands entirely destitute of wood. In fact, an incurious person might travel from one end of them to the other without seeing a single shrub. But in the ruts of streams, on lacustrine islets, occasionally along the shores of lakes, and in the clefts of rocks, there may be found stunted specimens of several species of trees. The common birch, the broad-leaved elm, the mountain ash, the hazel and the aspen are there casually met with. Willows of a few species are abundant along some of the rills, but seldom attain a height of three feet. *Rubus corylifolia*, *Rosa tomentosa*, *Lonicera periclymaenum* and *Hedera helix* are the only shrubs worth mentioning. 'Here is the observant botanist speaking. . . .'''

A memorial tribute to William MacGillivray was published (MacGillivray, 1901) within which some of his works are quoted. Being primarily an ornithologist, most of these quotations refer to birds, but there are passing descriptive botanical records. In an account of the Raven in the Hebrides, he says 'Even the water-lily, with its splendid white flowers, floating on the deep bog, reflects no glory on the surrounding scenery, but selfishly draws all your regard to itself.

(Brit. Birds, Vol. i p. 509)'. This intrinsically interesting statement does serve to demonstrate the nature of the botanical information in at least some of the early accounts.

A significant new phase in the exploration of the Outer Hebridean flora commenced in 1841 with the excursion by two recognised botanists, Professor John Hutton Balfour and Mr. Charles C. Babington, along with companions (Balfour & Babington, 1842 & 1844). They journeyed to Skye, North Uist, Harris and Lewis, and from the Outer Hebrides 'gave the number of plants for these islands as 311 species of flowering plants and higher cryptogams'. (Bennett, 1889). The 1844 publication is another source referred to by Campbell (op. cit.) as outstanding. Less well known is the Journal (Babington, 1897) which provides some detail of the 1841 journey, incorporating descriptive passages, anecdotal material and some botanical information. I have extracted the plant records from the Journal, retaining the Latin names as in the text, but relating the locations to those used in this volume (see end papers).

1841

Aug. 10. (19. North Uist) *Papaver argemone, Hippuris, Lycopsis arvensis, Salix herbacea.*

Aug. 12. (12 S. Harris) *Blysmus rufus.*

Aug. 13. & 14. (9 N. Harris) *Alchemilla alpina, Luzula spicata, Thalictrum alpinum, Saussurea alpina, Hymenophyllum wilsoni, Aira alpina.*

Aug. 16. (27. Shiant Is.) *Empetrum nigrum, Carex binervis, Lychnis dioica.*

Aug. 17. & 18. (1. Northern Lewis) *Lamium intermedium, Juncus balticus, Petasites vulgaris, Sagina maritima, Ranunculus sceleratus, Pyrethrum maritimum, Carex extensa.*

These are almost certainly first records for these species. It is interesting also to note that '. .people give the roots of the Petasites vulgaris, which abounds, to the cattle in winter'. This referred to Barvas, and has some bearing upon recent papers regarding the 'spread' of *Petasites hybridus* over the machair (Currie, 1977, Naylor & Cumming, 1892). Clearly, *Petasites* has been there for a very long time.

Another major phase in the exploration of the Outer Hebridean flora is shared throughout the British Isles, and can be identified with the work of H. C. Watson. One of the most significant pioneer works on the British Flora was Cybele Britannica (Watson, 1847–1859) (Cannon, 1978). Watson was a Yorkshireman who studied medicine in Edinburgh and came under the influence of Sir William Hooker. He frequently went on excursions with Hooker and his students, and wrote for over forty years on the geographical distribution of the British flora (Fletcher, op. cit.). Watson went on to prepare two volumes of 'Topographical Botany', published first in 1873 and 1874, with a second edition posthumously published in 1883. He created the 112 vice-counties into which the whole of Britain was divided, and of which the Outer Hebrides became V.C. 110. This had the effect of focussing attention on the island group as a geographical unit.

Arthur Bennett was a builder and house decorator, living in Croydon, England, who devoted a great deal of time and energy to the study of the Scottish flora. Between the years 1886 and 1911, he published a series of additional records to the 2nd edition of 'Topographical Botany' (1886 et seq., 1892b et seq.), including those for V.C. 110, and also a series of 'Contributions towards a Flora of the Outer Hebrides' (1892b et seq.). Many of the records were based on collections made for him by W. S. Duncan, who lived in the islands (Campbell, op. cit.). Fletcher (op. cit.) has commented that during all the years when Bennett was studying the Scottish flora and publishing records, he did not once visit Scotland.

Meantime, other botanists were also working. Peter Ewing (1890–1895) published 'A Contribution to the Topographical Botany of the West of Scotland' and in 1892 'The Glasgow Catalogue of Native and Established Plants'. Ewing revised his Catalogue in 1899 including records from the Outer Hebrides. In 1898, Professor James W. Trail of Aberdeen ömmenced the publication of a 'Topographical Botany of Scotland', followed by a series of additions and corrections which continued up to 1909. V.C. 110 featured in these. Somerville (1889 & 1890) and Shoolbred (1895 & 1899) are two other botanists whose publications were considered outstanding by Campbell (op. cit.), while Stirton (1885 & 1887) and Scott (1894 & 1895) also made contributions.

Nor were the outlying islands being overlooked. The paper by Barrington (1886) on St. Kilda was one of those commended by Campbell (op. cit.), while Gibson (1891) and Praeger (1897) also contributed to the lists for this most remote of all the outliers. A paper on St. Kilda by Mathieson (1928) included notes on the flora by Gladstone (1928). The plants which Gladstone collected went to W. B. Turrill, who published a paper (Turrill, 1928). Trail (1905) named a short list of plants collected on the Flannan Isles in 1904 by an ornithologist, W. Eagle Clarke, and similarly Bennett (1907) named additional species gathered in 1899 by W. J. Gibson. A paper on North Rona by J. A. Harvie-Brown included notes of the flora by Barrington (1885). Somerville (1891) exhibited plants from Mingulay near the southern extremity of the Long Island.

As a result of all of this botanical exploration, Bennett (1905b) was able to publish a 'Supplement to 'Topographical Botany' Ed. 2', incorporating records to that date. By 1929, Arthur Bennett, C. E. Salmon and J. R. Matthews had compiled a 'Second Supplement' which listed all records up to 1925, including those for V.C. 110. The final step in this early process of cataloguing the British Flora came when Druce published 'The Comital Flora' (1932). George Claridge Druce came from Northamptonshire and travelled all over the land for almost 60 years. In the Comital Flora, he gathered in one volume all the available information in what Dr. Fletcher called 'a modern version of Topographical Botany'. By this date therefore, the recorded flora of the Outer Hebrides was accessible within one volume.

## 3. 1930–1980

The period which commenced in the early thirties was significant in that a new approach to the study of the flora was evident. Early botanists named and listed plants, but the newer botanists related plants to their habitats and to vegetation communities. The account of the vegetation of St. Kilda (Petch, 1933) was the first 'ecological' account of the Outer Hebridean flora. This was followed by the scientific report of an expedition to Barra organised by the Biological Society of the University of Edinburgh (Forrest et al., 1936), which included an account of the vegetation (Watson & Barlow, 1936). In 'The Book of Barra' (Campbell, 1936) there is a list of plants, while there are later notes by Watson (1939) and Wilmott (1939). MacLeod (1949) who was a member of the 1935 expedition later revisited Barra and discussed plant ecology. Elton (1938) was another early ecologist, who provided notes on the vegetation of Pabbay in the Sound of Harris. Subsequent work on vegetation is fully dealt with by Pankhurst elsewhere in this volume.

A second significant new approach to the study of the flora became evident during this same period. Early botanists had based their accounts either upon brief visits or else upon information or specimens provided by others. The newer breed of botanist was prepared to go back year after year, recording on a more methodical basis, with a longer term view to writing a Flora based upon intimate knowledge of the Outer Hebrides. One of these prolonged studies was motivated by Professor J. W. Heslop Harrison. Richardson (1970) describes the origin of the study thus:– 'In 1935, as a result of joining an expedition from University College, Dundee, led by his friend Professor A. D. Peacock the previous year, he initiated a detailed ecological and biogeographical study of the islands of the Inner and Outer Hebrides. These studies carried out with parties of students and colleagues continued for some twenty years. . . .'. What was planned along with colleagues at King's College, University of Durham, was a comprehensive Flora of the Inner and Outer Hebrides. 'A Preliminary flora of the Outer Hebrides' was published (Harrison, 1941a) under Professor Harrison's editorship. This contained a list of 690 species, segregates in critical genera and hybrids of vascular plants. A bibliography of earlier publications by the Durham team was included, and the Flora was followed by a succession of further papers up to 1957, either adding to the Flora or dealing with specific localities.

A series of very useful general accounts of the aims, the progress and the conclusions of these studies may be found in Clark (1956), H. H. Harrison (1939), J. H. Harrison (1948) (reviewed Allen, 1951) and J. W. H. Harrison (1939b, 1941a, 1948b, 1950a & 1956b). Regrettably, the plan to prepare a comprehensive Flora was never achieved. There was also sadly a degree of controversy with regard to some records, making the work less valuable than it might otherwise have been. However many of the allegedly doubtful species have more recently been re-recorded, and the work of Heslop Harrison and his team remains a major contribution which cannot be ignored by any serious student of the Outer Hebridean flora.

The other prolonged study of the flora of the Outer Hebrides was that initiated by Miss M. S. Campbell. Her account of a visit to the islands in 1936 (Campbell, 1937a) was soon followed by another in 1937 (Campbell, 1938). In May 1938, Miss Campbell issued a printed appeal for assistance giving her address as c/o Department of Botany, British Museum (Nat. Hist.). The appeal was addressed 'To the Contributors to the New British Flora' and headed 'Flora of the Outer Hebrides'. She wrote 'During the past two years I have been working (with Mr Wilmott's co-operation) towards the publication of a Flora of the 'Outer Islands', concerning which there is a considerable scattered literature, but no comprehensive publication. The appeal is for help 'in determining doubtful identifications' and for 'notes on any Outer Hebrides material which you may examine in the course of your work'. The reverse side of the appeal has a useful bibliography.

Sadly, the war years interrupted the programme, though a few short botanical notes did appear between 1939 and 1945. Undoubtedly the most important publication to emerge from the whole study was 'The Flora of Uig (Lewis)' (Ed. Campbell, 1945) and subtitled 'A Botanical Exploration'. In addition to plant lists and systematic notes, this contained an account of the vegetation by Miss Campbell's close associate, A. J. Wilmott of the British Museum (Natural History). Wilmott published a further account of Uig following a post-war expedition arranged by Miss Campbell (Wilmott, 1948). These annual expeditions became a feature of the study. In an obituary (Milne-Redhead, 1984) the writer says 'Meanwhile another botanical interest had developed, with Wilmott's collaboration and Maybud's Scottish heritage, namely an annual botanising expedition to the Outer Hebrides, with a view to obtaining enough material and information to write a Flora. Other interested B.S.B.I. members used to join in, and the last pilgrimage to the Hebrides was as recent as 1981!' Something of the character of these expeditions may be read in the account of another field meeting led by Miss Campbell (Campbell, 1961), where the Lewis week is described by Miss U. K. Duncan and the Harris one by Dr F. H. Perring.

I myself met with Miss Campbell in Tarbert, Harris in 1976. She had sent volunteers out to botanise, and invited me to coffee in the Harris Hotel. We talked about her plans to write a series of Floras, beginning with one for Harris. Miss Campbell was later invited to write the chapter on the flora for the 1977 Symposium on the Natural Environment of the Outer Hebrides (Boyd, 1979). Unfortunately, she had to withdraw from the Symposium, feeling that the effort was too much considering the work that she still required to do on her own Flora. It was as a result of this withdrawal that I was invited to write the paper (Currie, 1979). The Provisional Checklist which was Appendix 1 of that paper was examined by Miss Campbell, who made useful comments, but the main support came from Dr F. H.Perring and the staff at the Biological Records Centre. Miss Campbell died in 1982 without having written a Flora, and in an eloquent obituary (Milne-Redhead, op cit.) the author commented 'The rich proceeds of these many expeditions are in the B. M. Herbarium, a fitting memorial to her love of Scottish botany, as is her 'Flora of Uig' (Campbell, 1945)'.

It would be futile to speculate what might have been achieved had these two separate teams felt able to cooperate. That there was a degree of enmity can be illustrated by reference to short papers by J. W.Heslop Harrison et al. (1938) and Campbell (1939b). A clash of personalities must have been a factor, one feels. That the potential for collaboration was there may be seen from the final sentence by J. W. Heslop Harrison (1948b) where he says 'Still, it must be admitted that much remains to be done, and by many workers; we give a hearty invitation to others to join us in our labours'.

Other B.S.B.I. field meetings took place in Stornoway (Copping, 1977), Barra (Conacher, 1980) and Stornoway once more (Currie, 1984, unpublished but the information incorporated in the present Flora). Quite apart from the work which went into recording on the main islands of the Outer Hebrides, a great deal of effort also went into visiting the less accessible islands and outliers. Miss Campbell visited several including Scalpay (Harris) (Campbell, 1944), while Harrison (1941a) gave a list which included Great and Little Bernera (Lewis), Pabbay (Harris), Berneray and Taransay (Harris), Baleshare, the Monach Isles, Eriskay, Fuday, Vatersay, Flodday, Muldoanich, Sandray, Pabbay (Barra), Mingulay and Berneray (Barra). Following the publication of the 1941 Preliminary Flora, some islands were revisited while new islands were also visited. In the following list, I have used for convenience the geographical zone numbers indicated in the map (see end papers).

 4. Great Bernera (Harrison, 1957)
11. Taransay (Harrison, 1954a)
12. Coppay (Harrison, 1954a)
13. Shillay (Harrison, 1954a)
14. Killegray (Harrison & Harrison, 1950b)
14. Ensay (Harrison & Harrison, 1950b)
17. Grimsay (Harrison & Harrison, 1950a)
17. Ronay (Harrison & Harrison, 1950a)
19. Stuley (Harrison & Harrison, 1950a)

It should be noted that many other small islands were visited, whose names appear within the text, but not in the titles, of the various papers. The above list, however incomplete, does however give some idea of the geographical extent of the effort. Many other botanists have also visited the remote and outlying islands during the period under discussion. See the appendix for the bibliography of the flora of particular islands. There have of course been many other visits which were either not documented or else which were presented in an ephemeral form. For example, many schools, colleges and environmental groups have spent time in remote islands, producing for their own use reports which contain botanical information which remains rather inaccessible.

In concluding this section of the history of the study of the flora of the Outer Hebrides, it is particularly worthwhile to comment upon some published works which have a bearing upon plant conservation or habitat management, and whose authors were conservationists. Early works (Darling, 1947 & Darling & Boyd, 1964) were of a general nature, but contained a great deal of botanical information from the Outer Hebrides. Specific to St. Kilda, the study of the Soay sheep (Jewell, Milner & Boyd, 1974) similarly contains information about the botany, including plant communities and a vegetation map. Ratcliffe (1977a) touches on the flora of the Outer Hebrides within the Highland context. A series of three reports on sand dune machair (Ranwell, 1974, 1977 & 1980) contain a great deal of additional information on this particular botanical habitat.

In a major review of the selection of biological sites of national importance to nature conservation in Britain (Ratcliffe, 1977b), Outer Hebridean sites are discussed in their national context with regard to botany among other aspects. In particular, Vol. 2 provides site accounts for individual sites, within many of which is included botanical information. Those with botanical data are listed under habitats below.

*1) coastlands*

South Uist Machair (a) Grogarry (b) Askernish coast
Balranald, North Uist
Monach Isles
St. Kilda
Baleshare/Kirkibost Dunes, North Uist

*2) open waters*

Loch an Duin, North Uist
Grogarry Lochs, South Uist (a) Loch Druidibeg (b) Loch a'Mhachair and Loch Stilligarry (c) Howmore Estuary, Loch Roag and Loch Fada
Loch nam Feithean Balranald, North Uist

*3) peatlands*

Little Loch Roag Valley Mire, Lewis

*4) upland grasslands and heaths*

North Harris

It should be noted that these sites are constantly being reviewed, and other sites added. See the appendix of National Nature Reserves and Sites of Special Scientific Interest.

On 11th and 12th October, 1977, a Symposium on the Natural Environment of the Outer Hebrides, organised by the Royal Society of Edinburgh and the Nature Conservancy Council, was held in the Society's rooms in Edinburgh. The object of the Symposium was to bring together as many as possible of the current works in the field, both in review and in original papers, and therefore provide a body of environmental information. (Boyd, 1979b). As well as papers describing the physical environment of climate, geology and soils, there were several botanical papers. These dealt with vegetation, including a provisional checklist (Currie, 1979), machair vegetation (Dickinson & Randall, 1979), peatland vegetation (Goode & Lindsay 1979) and macrophytic vegetation of fresh and brackish waters (Spence, Allen & Fraser, 1979). Boyd (1979c) provided lists of four National Nature Reserves and 39 Sites of Special Scientific Interest, many though not all of which have been declared for their botanical interest. These S.S.S.I.s are constantly under review, and while some may have been dropped, others may have been extended and new sites added. Botany can only benefit from this constant management and review of plant habitats.

## 4. THE 1980s

In considering the state of botany in the Outer Hebrides in the 1980s, it is interesting to recall how the aims of botanical study have changed. In the very early years, people were really exploring; then came the early botanists who identified what they saw and made lists; later came the ecologists who related plants to their environment; by the 1980s conservation of the flora came to the fore. During this final period, it was an anticipated threat which stimulated a great deal of intense research into both fauna and flora of the Outer Hebrides. 'An Integrated Development Programme for the Western Isles of Scotland (Outer Hebrides)' was a five year programme 1982–87. It arose from Council Regulation E.E.C. No. 1939/81. The basic objective of the I.D.P. as set out in that Regulation was to 'improve working and living conditions in the Western Isles' through a series of measures designed'....to improve agriculture; to improve the marketing of agricultural (and fisheries) products – including the afforestation of marginal land, operations to improve the marketing and processing of agricultural products and measures to develop fisheries, but also measures relating to tourist amenities, crafts, industrial and other complementary activities essential to the improvement of the general socio-economic situation of those isles'. There was also an environmental objective – to maintain a proper balance between human needs and the natural environment.

Many saw the proposed I.D.P. as a threat to the natural environment, and in particular the machair with its very high botanical and ornithological interest. The result was that a great deal of effort went into recording the habitats seen to be under threat, and the material collated within the Symposium volume (Boyd, 1979b) proved to be of immense value. The Nature Conservancy Council is the body responsible for advising the Government on nature conservation in Great Britain, and it took a lead in providing the considerable resources required for specialist survey work. The reports produced to date are internal documents, and only those referring to botany need be mentioned here. Pitkin et al. (1983) carried out botanical survey of S.S.S.I.s in the Uists, while Mackintosh & Urquhart (1983) surveyed haymeadows also in the Uists. In an important commissioned survey, the Royal Botanic Garden, Edinburgh (1983, 1984 & 1985) surveyed aquatic vegetation and freshwater macrophytes in the Uists, Lewis and Harris, covering a great many lochs. N.C.C. Edinburgh (1986) is the report of the environmental appraisal, Part 3 of which contains the reports of specialist surveys. The surveys included a major input by voluntary conservation bodies, including a habitat survey of the Uists carried out by the Scottish Wildlife Trust (Philp, 1983). The botanical information from all of these studies was fortunately available to the writers of the present Flora. In the end, the damage done to botanical habitats by the I.D.P. was minimal, but it does demonstrate the need for all botanists to be alert to future threats to the environment, even in more remote locations such as the Outer Hebrides. It is to be hoped that the new botanical information acquired will soon be published in more accessible journals.

# Plant Lore in Gaelic Scotland

Margaret Bennett

*School of Scottish Studies, University of Edinburgh, 27, George Square,* Edinburgh EH8 9LD

Amid the ruins on the most remote Highland croft still remains ample evidence of age-old customs and beliefs concerning plants and their significance in the daily lives of the people who lived there. The rowan tree (*Sorbus aucuparia*) or **Caorann** in Gaelic, which stands in the corner of the kitchen garden or just beyond the house or byre, has protected the home, the family and the cattle from witches and fairies for centuries, and has endured after many a homestead has been deserted. No Highlander worth his salt would cut one down, nor will he allow his son to do so, even if it means walking many a long stretch to fetch home firewood. Belief in the power of such plants can be traced back to the time of the Druids; it was certainly part of the belief system of the Celtic peoples, and although it may not be voiced so explicitly as it once was, yet the custom of planting and preserving this special tree is still continued by some people. Nowadays it is merely regarded as good luck.

In his book *The Gaelic Names of Plants* (1883) John Cameron notes that the 'Celts named plants often from (1), their uses; (2), their appearance; (3), their habitats; (4), their superstitious associations, &c' and to clarify some of the points he wishes to make he notes comparisons between Scots, Irish and Welsh names. Martin Martin (1703) who wrote his well-documented *Description of the Western Islands of Scotland* has left a wide and varied record of many of the plant uses which were at that time extant. The oldest book source that Highlanders might refer to for plant lore is probably the Bible. We need not, however, rely solely upon the works of old masters to bring to mind the lore that is associated with numerous plants. For example, many a small child knows that to alleviate the pain and irritation of a nettle sting he need only reach for the nearest broad-leaved docken, **Copag Leathann** (*Rumex obtusifolius*) and rub it on the offending part for instant relief. Cameron's four headings are, however, useful broad categories into which might fall the examples chosen for this paper.

The uses of plants are legion: for medicine, for food and drink, for domestic, agricultural and seafaring purposes, for magical functions, or simply for luck. To return to the offending nettle (*Urtica dioica*) or **Deanntag**, it too had its uses: Martin Martin noted that in Skye 'the tops of nettles, chopped small, and mixed with a few whites or raw eggs, applied to the forehead and temples by way of a frontel, is used to procure sleep'. He also recorded in Lewis a home-remedy which used the roots of nettles boiled in water along with the roots of reeds, then fermented by the addition of yeast which 'they find beneficial for the cough'. Perhaps more convincingly, the nettle tips were (and still are) a spring-time green vegetable, high in vitamins and minerals, a valuable supplement to diet. Nettle tea, simply made by infusing in boiling water, and nettle soup, were also very common. On several of the islands the entire plant, boiled in water, was used for dying wool a greenish-yellow colour.

Plants as a supplement for diet were not only important but essential. Highly valued among a people who claimed its virtues in several ancient songs was **Brisgean**, silverweed, (*Potentilla anserina*):

**Brisgean beannaichte earraich,**
**Seachdamh aran a' Ghàidheil.**

[The blest silverweed of spring,
*One of the seven breads of the Gael.]

Before the introduction of the potato the versatile root of the silverweed was commonly boiled in water, roasted on the fire, or dried and ground into meal for bread-making or porridge. It was especially welcomed by the poor Islanders who were cleared from their crofts

---

* Literally: 'A seventh part of the Gael's bread'

during the infamous Highland Clearances, and at the time of the potato blight many were said to have subsisted on it.

White or pink stonecrop (*Sedum anglicum*) was considered to be a delicacy and was given the name **Biadh an t-Sionnaidh**, the prince's or lord's food, also known as the English stonecrop. Even nowadays children at play can still be seen eating the occasional leaf of sorrel (*Rumex acetosa*) which is generally called **Sealbhag**, possibly from the word for 'sour', **searbh**. Their forebears may well have relied heavily upon scurvy grass (*Cochlearia officinalis*) in some areas referred to as **Maraiche**, 'a sailor' or 'mariner', and in other-areas called **Carran**, 'the thing for scurvy', or more explicitly, possessing antiscorbutic properties. It is already widely-known that sailors suffered the dreaded disease of scurvy (**carr** in Gaelic), which could be counteracted by eating the food containing the appropriate substance, ascorbic acid. On land, and in particular in the Highlands and Islands of Scotland, this was certainly not in the well-known form of lime juice, as the Gaels applied their own wisdom for the prevention of the disease. Martin Martin notes that on a small rocky island to the south of Skye there is 'a great quantity of scurvey-grass, of an extraordinary size, and very thick; the natives eat it frequently, as well boiled as raw: two of them told me that they happened to be confined there for the space of thirty hours, by a contrary wind; and being without victuals, fell to eating this scurvey-grass, and finding it of a sweet taste, far different from the land scurvey-grass, they ate a large basketful of it, which did abundantly satisfy their appetites until their return home'.

From time immemorial it was a commonly held belief in Gaelic Scotland that where a disease occurred there would be found growing in the locality of its prevalence a plant which would cure that disease. This idea has raised many an eyebrow, especially during our modern age, yet in 1983 the World Health Organisation launched a ten-year study into herbal remedies. The intention is that doctors will know which herbs can be used in the treatment of certain complaints. According to Dr. Desmond Corrigan of Dublin University, 'it is a fact that 40% of all drugs used at present have a folklore basis'. The use of the foxglove (*Digitalis purpurea*) or **Lus nam Ban-sìdh**, 'the fairy women's plant' in Dr. William Withering's 18th century discovery of the drug digitalis is well documented. It is still one of the most commonly prescribed drugs for heart complaints. Dr. Corrigan suggests that 'by the time the World Health Organisation have listed and checked all the herbs which could be used, doctors may find themselves prescribing herbs which were once dismissed as old wives' tales'.

In view of this recent move by the W.H.O., the work of Martin Martin has a large contribution to make. To cite but a few examples, he notes that a 'refreshing drink for such as are ill of fevers' was whey in which was boiled violets (*Viola riviniana*), **Dail-chuach**, 'field bowl'. Kidney disorders, or the 'stone', as such ailments were called, were generally treated with various infusions such as those made with scurvy-grass or wild garlic (*Allium ursinum*), **Gairleag** in everyday language, and **Creamh** in poetry. It was also known as **Gairgean**, which has a second meaning, 'an irritable person'. The plant was boiled in water and the resulting infusion was said to be an effectual diuretic.

Martin recorded several plants which indicated diuretic properties, some of which had purgative effects, and others which 'cured the fluxes', such as the syrup of bilberries. His observations made in 1695 are not only an important source of early medicine (and therefore of world significance) but are also a vital source of Hebridean social history. We can now observe twentieth century features of Hebridean kitchen gardens and can recognise a profusion of plants which occur, then sense the continuity of custom as Martin Martin informs us of the uses he observed almost three centuries ago. This is no mere coincidence but an indication of the antiquity of knowledge among the Gaels who cultivated these plants close to almost every house they built. Two very common examples from the same plant family are feverfew and tansy. The former (*Tanacetum parthenium*) is called **Meadh Duach** in Gaelic and was used as a cure for migraines, while the latter (*Tanacetum vulgare*), **Lus na Frainge**, (literally translated 'the French weed') had a rather different use which was recorded by Martin: 'To kill worms, the infusion of tansy in whey, or aquavitae, taken fasting, is an ordinary medicine with the islanders'. If the taste or effect of taking tansy proved too daunting a thought, perhaps the patient might have consoled himself that the suspension of aquavitae could have aided the effects of the medicine if not improved the taste.

In the making of aquavitae certain plants were utilised as additives, such as some members of the Leguminosae or vetch family, **Peasair nan Luch**, literally 'mice pease' and known simply as vetch, and **Cairmeal** or **Cairt Leamhna** known as bitter vetch (*Lathyrus montanus*). Because of its aromatic properties it was preferred to spice in the making of aquavitae. It had other beneficial and related properties too, as Martin notes that 'the plant itself is not used, but the root is eaten to expel wind, and they say it prevents drunkenness by frequent chewing of it'. He adds that 'the natives of Mull are very careful to chew a piece . . . especially when they intend to have a drinking bout'. Better an ounce of prevention than a pound of cure!

Sundew (*Drosera rotundifolia*) has, according to one interpretation of the Gaelic, a somewhat uncomplimentary name – **Lus na Feàrnaich**, said to be 'the plant of **earnach**', a disease in cattle, sometimes identified as murrain, and reputed to be caused by eating this poisonous plant. It had some redeeming features, however, as it was used, according to Cameron, as a beauty aid 'much employed among Celtic tribes for dyeing the hair'. A second interpretation of its Gaelic name is 'plant with the shields', a more picturesque one attributed to the shape of the sundew's leaves.

Better known and more commonly employed in the daily lives of the Hebridean people were the plants used for dyeing wool. The sundew was used in the making of tartan to produce a fine purple, but best known of all is **crotal** (*Parmelia saxatilis*), a kind of lichen whose Gaelic name has gained more popular currency among English-speakers in Scotland than the equivalent translation, stone parmelia. It dyes wool a reddish brown colour, distinctively well known and often referred to in either language as **crotal**. The lichen is gathered from rocks late in the summer, dried in the sun, then it is placed with the wool in alternating layers in a large pot. It is then covered with water, placed on a fire, and simmered till the desirable depth of colour is obtained. Finally the wool is removed, rinsed in water with salt added to set the dye, and dried in the sun. Attached to this plant and the dye it produces are several beliefs which are especially pertinent to the families of fishermen. These men will not readily wear garment of crotal when going to sea, as it is said that the crotal plant was plucked from the rocks and will therefore return to the rocks. It is also believed that if a person drowns wearing crotal the body will never be recovered.

Recipes for plant dyes tested and tried over generations have produced distinctive colours for wool, tweed and tartan. Some plants are used in their entirety, such as bog myrtle (*Myrica gale*), **Roid**, which produces a fine yellow, or nettle which has already been described. Others use only the root, such as the yellow flag or iris (*Iris pseudacorus*), **Sealasdair** or **Seileasdair**, which is dug up after the flower has bloomed. It is then cleaned, cut up and boiled in water to produce a blue-grey. The roots of lady's bedstraw (*Galium verum*), **Rùin**, produce an orange-red colour which is set by adding alum. The flowers of the corn marigold (*Chrysanthemum segetum*), **Bile Buidhe**, gathered in a muslin bag and boiled in water with alum, produce a bright, clear yellow. The tops of ling heather (*Calluna vulgaris*), **Fraoch**, produce yet another shade of yellow dye when boiled in water, and in fact that entire plant is wonderfully versatile: it was commonly used for thatching houses, and even today the few Highland thatchers that remain will swear it is the best thatch in Gaeldom; it provided beds to sleep on, with the 'tops up and roots down' arrangement of the mattress assuring a pleasantly aromatic and sound sleep; it was used in part of the process of tanning leather; and the fresh, young tops of the heather were (and at times still are) brewed into a kind of ale. Little wonder it is acclaimed in song and story and longed for by the expatriate Scot!

Country folk are well known to observe signs which indicate the weather. Some of the signs, such as the quantity of berries on certain trees in autumn were for long-range forecasts, while still others could tell the weather for that very hour: '**Tha'n t-seamrag a' pasgadh a còmhdaich ro thuiteam dòirteach**' (The Shamrock is folding its garments before a heavy downpour) is not only a picturesque saying but tends also to be accurate.

Certain plants have Gaelic names which assign particular qualities to them, such as **Lus an Leanna**, 'the beer plant', which is commonly known as hop(s) (*Humulus lupulus*). The product used was made by boiling a quantity of hops with treacle, adding mashed potatoes and salt, then thickening the mixture with flour. It was later stored in a crock and used in baking. Yarrow (*Achillea millefolium*) has one Gaelic name **lus chasgadh na fala**, 'the plant that staunches bleeding', as it is reputed to contain an element which aids the clotting of blood. This plant has been recorded in the writings of ancient Greeks and Romans who noted that it was especially efficacious in 'stopping the blood of a wound inflicted by iron', presumably in battle. Yarrow is also known as **Eàrr-thalmhainn**.

One of the Gaelic names for poppy is **lus a'chadail**, 'the sleep plant', which is probably *Papaver somniferum* because of the acknowledged soporific properties. The most common poppy in the Outer Hebrides is *Papaver dubium* whose beautiful red blooms brighten the early summer fields.

The shape or general appearance of many plants gives rise to numerous names which have appeal to the senses of sight and smell as well as the imagination. The mountain everlasting (*Antennaria dioica*) with its dainty little furry flowers is known in Gaelic as **Spòg Cait**, literally 'cat's paw', while the marsh marigold (*Caltha palustris*) is known as **Lus Buidhe Bealltainn**, 'the yellow plant of Beltane', for it makes its appearance at Beltane, May first, thus marking the second half of the ancient Celtic year. Red campion (*Silene dioica*) with its bright red jagged flowers is aptly known as **Cirean Coilich**, 'cock's comb', while ragged robin *Lychnis flos-cuculi* is referred to as **Caorag Lèana**, giving the bright picture of a 'marsh spark'. The Scottish bluebell (*Campanula rotundifolia*) is sometimes called **currac na cuthaig** 'the cuckoo's hood', a beautiful image for such an elegant flower. The greater plantain (*Plantago major*) is known as **Cuach Phàdraig**, [Saint] 'Patrick's bowl'. Hemlock (*Conium maculatum*) has the Gaelic name of **Iteodha**, which it is suggested by Cameron, is possibly from the word for feathers, **ite**. Although there are also other Gaelic names for hemlock, the Biblical reference in Hosea X. 4 uses this name: **'Mar seo tha breitheanas a' fàs a-nìos, mar an iteotha ann an claisean na machrach'**. (Thus judgment springeth up like a hemlock in the furrows of the field). Hemlock water-dropwort (*Oenanthe crocata*), however, has a Gaelic name that suggests another kind of imagery: **fealladh bog**, meaning 'gentle deceit'. It was said to have been given to prisoners as poison from the time of Pliny, and most references agree that it was also the plant with which Socrates was poisoned. In the Outer Hebrides there is, however, a second and more urgent suggestion of deceit, as this poisonous plant looks very much like water-cress, and therefore one could be deceived into eating it, with disastrous results. And on another cautionary note, there is the white water lily (*Nymphaea alba*), which goes by name of **Duilleag-bhàite Bhàn**, 'the white leaf of drowning' ... its extreme beauty may lure the admirer to a watery grave. The common fumitory (*Fumaria officinalis*) is known in Gaelic as **Lus Deathach-thalmhainn**, which translates as 'earth-smoke plant' and is a play on the Latin name. It was believed that the smoke of this species had the power of expelling evil spirits.

There were several plants which were significant to generations of Hebridean children who were brought up to have a knowledge of them. Aside from the cautionary tales that accompanied some of the Gaelic plant names, there were some plants that were the focus of childhood games. An almost universal game was 'Soldiers' played with a stalk of plantain: **'Thugainn a' chluich saighdeirean'** ['Let's play soldier's'] was the usual invitation to play. Two children (the combatants) each chose a 'soldier' and held battle until one managed to chop off the head of his opponent. Often the winner could go into battle many times with the one sturdy 'soldier' as there were some stems of the plantain which (according to a veteran Uist player) were 'like grizzled warriors and would knock out hundreds!'

Another plant used in a most imaginative manner by children at play was **buaghallan,** ragwort (*Senecio jacobaea*) which was the central object of a game called **'Goid a' Chruin'** ['Steal the Crown']. It was a popular game on several of the Hebridean islands and though there were regional variations, this version is from the Rev. N. MacDonald who was recorded by the late Calum Iain MacLean of the school of Scottish Studies in 1953. Originally in Gaelic (and here translated) it gives a general account of the game which was played as follows:

'Well, we played it during the interval at school when I was a little boy. You had always to be on a level piece of ground, you see, with no rise whatsoever and we always had that; we had an excellent stretch of green sward a short distance from the school, and we gathered there. And it was always in autumn that this game was played because we had a sort of a plant that is very common then, we call it **buaghallan** in Gaelic [ragwort]. And that was plucked out by its roots, and two rows were set up on the green; boys and girls were in each row. **An Crun** [The Crown] was one of those **buaghallans**, one of those plants. It was thrown there on neutral territory between the two ranks and the person who was successful in breaking through and taking up the crown and running right down ... round, circling their opponents, and right round and coming back without being caught, that person was successful. But he always had to give the next, his companion, a chance, and he himself just looked on. But if he was caught with the **buaghallan,** with the crown, by his opponents before he circled them and came back to his own allies . . . then he had to go outside the camp, and he was called a **cnoimheag** [maggot!]. Now the side that had more failures (or more **cnoimhags**) was the side that lost, and the row that had less, of course, was the one that was successful. And that was the game!'

A few plants are associated with historical lore such as the pink convolvulus or sea bindweed (*Calystegia soldanella*), which is called **Flùr a' Phrionnsa**, 'the Prince's flower', because it was said to have been originally sown by Prince Charles in 1745 when he landed in Eriskay, and it is still growing on that island.

As stated earlier, certain plants were said to have supernatural powers, often revealing an interesting combination of Christian and pagan beliefs in the one element. Slender St. John's-wort (*Hypericum pulchrum*), for example, was in some areas called **caod achlasan Chaluim Chille**, 'the flower carried in the arms of St. Columba'. In pre-Christian times it was carried as a charm against witchcraft, a notion which survived long after the arrival of Christianity. In *Carmina Gadelica* (1900), Alexander Carmichael notes that 'St John's wort is one of the few plants still cherished by the people to ward away second-sight, enchantment, witchcraft, evil eye, and death, and to ensure peace and plenty in the house, increase and prosperity in the field, and growth and fruition in the field'. It was said to be effective only if found without actually seeking it, and some people who came upon it used to say a prayer such as:

**Achlasan Chaluim Chille,**
**Gun sireadh, gun iarraidh!**
**Dheòin Dhia agus Chriosda**
**Am bliadhna chan fhaigheas bàs.**

Saint John's wort, Saint John's wort,
Without search, without seeking!
Please God and Christ Jesus
This year I shall not die.

Carmichael states that 'the plant is secretly secured in the bodices of the women and in the vests of the men, under the left armpit', while Martin Martin gives an account of a man in Berneray, Harris, who wore it in the neck of his coat to prevent him from seeing visions, and 'he never saw any since he first carried that plant about with him'. It was also said to possess curative powers, as did several other plants which incorporate the name of the Virgin Mary or the saints in their Gaelic names.

Clan badges are many and varied, and several have been selected from the local plants. To cite but a few examples, the thistle is the official badge of the Clan Stewart. With several species in the Outer Hebrides such as **Cluaran Lèana** ( *Cirsium palustre*) the marsh thistle, **Cluaran Deilgneach** (*Cirsium vulgare*), the spear thistle, and **Fòthannan Achaidh** (*Cirsium arvense*), the field thistle, a prospective wearer need not be short of choice. Bog myrtle (*Myrica gale*) **Roid**, sometimes rendered as **miortal**, is the badge of the Clan MacArthur; and the MacDonalds and MacAlisters have for their clan badges the heather, already mentioned for its great versatility.

The Gaelic alphabet is represented by eighteen trees, beginning with *A*, **ailm**, elm ( *Ulmus*) and ending with *U*, **uir**, hawthorn *Crataegus monogyna* . . not quite an A–to–Z of plants, for there is no Z in Gaelic. Represented here is a mere sampler from an enormous tapestry of Gaelic plant names and their associated lore.

# Plan of the Flora of the Outer Hebrides

Each of the species accounts is laid out as follows:
- Scientific name, as in Clapham, Tutin & Moore.
- English name (from the BSBI standard list).
- Gaelic name, if known, from list compiled by J. W. Clark & I. MacDonald, see below.
- Status.
- Habitat, especially if different from mainland.
- First record (introductions only).
- Distribution, as list of zones by numbers, with comments if appropriate. The zone numbers are given in brackets if the record is doubtful. More precision is given for rarer plants. In such cases, we may have cited; collector's name (see Appendix) in *italic* and date, author(s) and date in normal type for literature records e.g. Harrison (1941a), see appendix for bibliography, the herbarium code name for voucher specimens.

In many cases, we have not had enough information to be able to give separate distributions for intraspecific taxa.
- Taxonomic notes, including short keys as and when appropriate.


*Notes on the Gaelic names*

The main published sources of Gaelic plant names are John Cameron's out of print compilation *The Gaelic Names of Plants* (Blackwood and Sons, Edinburgh and London, 1883; 2nd ed., John Mackay, Glasgow, 1900) and Edward Dwelly's *The Illustrated Gaelic–English Dictionary* (1901–11, and often reprinted, most recently in 1988 by Gairm Publications, Glasgow). Dwelly in fact lists many of the plant names in Cameron, on occasion presenting corrected forms of them, but also draws on other sources which he cites at the front of his dictionary. Names of plants will also be found in a collection such as *Gaelic Words and Expressions from South Uist and Eriskay*, collected by Father Allan McDonald and edited by J.L.Campbell (2nd ed., Oxford University Press, 1972). More recently, a short list has ·appeared in a Gaelic work on biology by MacLeòid and MacThómais, *Bith-eòlas* (Gairm Publications, 1976), while a certain number can be found in dictionaries, old and recent, other than Dwelly's. There are several quite elaborate private lists in existence, but none of these has yet been published.

As might be expected when the large area in which Gaelic was widespread is considered, plant names may differ from place to place or be recorded in variant forms, and the choice (in most cases) of one name only in the Flora was extremely difficult. Published and unpublished lists were used in compiling the necessarily limited but, it is hoped, representative selection of Gaelic names that appears here, and help was given by Gaelic speakers in the Outer Hebrides and elsewhere. Systematic investigation of the authenticity of the names and of their etymology has not been attempted, although the orthography has been brought up-to-date. This does not, therefore, purport to be a standard list, and it is recognised that an up to date and authoritative list is greatly to be desired.

In many instances Dwelly and others provide only the plant's generic name, and where necessary the different species have been distinguished by qualifiers such as *lòin* ('meadow'), *lèana* ('marsh') etc. Where * appears, the generic name has been introduced recently and is not to be found in the older sources, while names marked ** are based on Irish exemplars, no Scottish names being known, and have been adapted for Gaelic as necessary.

J. W. Clark & I. MacDonald.

**PTERIDOPHYTA**
*LYCOPODIACEAE*
**Lycopodium** L.
**L. clavatum** L.
Stag's-horn Clubmoss.
Lus a' Mhadaidh-Ruaidh, Garbhag nan Gleann.
Native.
Rare, in the Stornoway area of Lewis only; bare heath East of Stornoway *Lt.Col. Macrae* 1938 (BM); Castle Grounds *Miss Macarthur* comm. *Lt.Col. Macrae* 1939 (BM); heath on road to Melbost Farm (perhaps the same as first) *Lt.Col. Macrae* 1939 (BM)    1, 5

**Lycopodiella** J. Holub
**L. inundata** (L.) J. Holub
*Lycopodium inundatum* L.
Marsh Clubmoss.
Garbhag Lèana.
Native.
Only one record, Harris; Loch an Tairbh Duinn near Tarbert *Burnier* comm. *Ounstead* 1968 see Watsonia 8:53 1970    12

**Diphasiastrum** J. Holub
**D. alpinum** (L.) J. Holub
*Lycopodium alpinum* L.; *Diphasium alpinum* (L.) Rothm.
Alpine Clubmoss.
Garbhag Ailpeach.
Native.
North Harris; Mullach an Langa, in 1841 (*Balfour and Babington*, 1844); between Sgurr Scaladale and Tomnaval *Bowman and Bowman* 1977 (LTR); also flank of Tomnaval (*Bowman and Bowman* 1979).    9

**Huperzia** Bernh.
**H. selago** (L.) Bernh. ex Schrank & Mart.
*Lycopodium selago* L.
Fir Clubmoss.
Garbhag an t-Slèibhe.
Native.
Mountain grassland and loch margins.
South Uist to Lewis, St Kilda    3–5, 7, 9, 12, 13, 15, 16, 18, 19, 21, 24, 28

*SELAGINELLACEAE*
**Selaginella** Beauv.
**S. selaginoides** (L.) Link
Lesser Clubmoss.
Garbhag Bheag.
Native.
Machair, flushes and damp grassland.
Berneray and Mingulay to Lewis and St Kilda.    1–3, 6, 7, 9, 12, 13, 15–19, 21–24, 28, 29

*ISOETACEAE*
**Isoetes** L.
**I. lacustris** L.
Quillwort.
Luibh nan Cleiteagan.
Native.
Oligotrophic lochs.
Vatersay to Lewis    3, 4, 12, 16, 18, 19, 21, 22

**I. echinospora** Durieu
*I. setacea* auct.
Spring Quillwort.
Luibh nan Cleiteagan.
Native.
Oligotrophic lochs.
Lewis, South Uist to Harris, Scarp    1, 8, 12, 18, 19

*EQUISETACEAE*
**Equisetum** L.
**E. × trachyodon** A. Br.
(*E. hyemale* L. × *E. variegatum* Schleider ex Weber & Mohr)
A specimen from Nisibost, South Harris, originally named as *E. variegatum* has been redetermined by *Page* as this hybrid, see Murray (1981), and Page & Barker (1985). Also recorded from Berneray, (Pitkin et al., 1983).    12, 15

**E. variegatum** Schleicher ex Weber & Mohr
Variegated Horsetail.
Native.
Lewis, harbour at Skegirsta, Harrison (1957); South Harris at Scarista, Ben Capval, Luskentyre and Nisibost; North Uist, dunes at Eachkamish, *Murray* 1977. (see entry for *E. X trachyodon*).    1, 12, 16

**E. fluviatile** L.
Water Horsetail.
Clois.
Native.
Shallow lochs and marshes.
Barra to Lewis    1–3, 5–7, 9, 12, 16–22

**E. × dycei** C. N. Page
(*E. fluviatile* L. × palustre L.)
Loch a' Mhorghain, North Harris *Page* 1962 (E, there is also a photograph and fragment at BM). See Page, 1963 (illustration) and Page, 1981.    9

**E. × litorale** Kühlew. ex Rupr.
(*E. arvense* L. × fluviatile L.)
South Uist to Lewis and St Kilda    1, 6, 12, 19, 28

**E. palustre** L.
Marsh Horsetail.
Cuiridin.
Native.
Marshes.
Mingulay to Lewis and St Kilda   1–6, 9,
12–16, 18, 19, 21–24, 28

**E. sylvaticum** L.
Wood Horsetail.
Cuiridin Coille.
Native.
Marshes, flushes.   Mingulay to Lewis 1, 3,
5, 6, 8, 9, 12, 16, 21, 24

**E. pratense** Ehrh.
Shady Horsetail.
Native.
Recorded from Obbe, South Harris *W. S.
Duncan* 1889 (BM); foot of Lee Hills, North
Uist in 1898 (*Shoolbred* 1899); Loch Ollay,
South Uist *J. W. H. Harrison* et al., 1939
and 1941a, but no specimens have been
seen.   12, ?16, ?19

**E. arvense** L.
Field Horsetail.
Earball an Eich.
Native.
Riversides, loch margins, ditches and
machair.
Barra to Lewis and St Kilda   1, 3–6, 9,
12–16, 18–24, 28

**E. telmateia** Ehrh.
Great Horsetail.
Earball an Eich Mòr.
Native.
Near West Loch Tarbert, Harris *Page* 1962
(BM) det *Hauke* 1975

*OPHIOGLOSSACEAE*
**Botrychium** Swartz
**B. lunaria** (L.) Swartz
Moonwort.
Lus nam Mìos, Luan-lus.
Native.
Grassy slopes and machair.
Barra to Lewis and St Kilda   3, 4, 12–16, 18,
19, 21, 28

**Ophioglossum** L.
**O. vulgatum** L.
Adder's-tongue.
Teanga na Nathrach.
Native.
Grassy places and machair.

Mingulay to Lewis and Monach Is. to St
Kilda and North Rona   1, 8, 12, 13, 15–22,
24, 25, 28, 29

**O. azoricum** C. Presl
*O. vulgatum* L. subsp. *ambiguum* (Cosson &
Germ.) E. F. Warburg
Specimens from St Kilda *Barrington* 1883
(BM), Scarp *W. S. Duncan* 1890 (BM) and
South Uist *Cleave* 1983 and literature records
from Benbecula, Scotasay and North Rona.
Paul (1987) concludes that this species is
possibly only an extreme form of *O. vul-
gatum*.   8, 10?, 18?, 19, 25?, 28

*OSMUNDACEAE*
**Osmunda** L.
**O. regalis** L.
Royal Fern.
Raineach Rìoghail.
Native.
Streams and loch margins and sea cliffs.
Quite frequent, in contrast to the Shetlands,
where it is a rarity, said to be on account of
grazing by sheep (Scott and Palmer, 1987).
Sheep are probably just as ubiquitous and
destructive in both places, so this cannot be
the complete explanation.
Mingulay to Lewis   1–5, 7, 8, 10–13,
16–19, 21, 23, 24

*ADIANTACEAE*
[*SINOPTERIDACEAE*]
**Cryptogramma** R. Br. ex Richardson
**C. crispa** (L.) R. Br. in Hooker
Parsley Fern.
Raineach Pheirsill.
Native.
Loose rock scree.
Near Rodel, South Harris, 1841, (*Balfour
and Babington* 1844); Strone Scourst, North
Harris *M.S. Campbell* 1938 (BM); Beinn
Mhor, South Uist (*J. W. H. Harrison*, 1941a);
Clisham, North Harris (ibid.); Great Bernera,
field record *Currie*, 1984.   4, 9, 12, 19

*MARSILEACEAE*
**Pilularia** L.
**P. globulifera** L.
Pillwort.
Feur a' Phiobair.
Native.
Rare. South Harris, Obbe, *W. S. Duncan* 1891
(BM); South Uist, Loch Kearsinish, *Preston*,
1987 (BM).   12, 19

*HYMENOPHYLLACEAE*
**Hymenophyllum** Sm.
**H. wilsonii** Hooker
Wilson's Filmy-fern.
Raineach Còinnich.*
Native.
Wet rocks and mountain ledges.
Wilmott (1948) remarks on the remarkable association of this species with *Salix herbacea* on the summit of Suainaval (Lewis).
Vatersay to Lewis and St Kilda   3, 4, 7, 9, 11, 12, 16, 18, 19, 21, 22, 28

*POLYPODIACEAE*
**Polypodium** L.
**P. vulgare** L.
Polypody.
Clach-raineach.
Native.
Rocks and walls.
Mingulay to Lewis and St Kilda.   3–5, 7, 9, 10, 12, 13, 16–18, 19, 21–24, 28

**P. interjectum** Shivas
Clach-raineach.
Native.
Rocks.
First collected from Borosdale near Rodel, South Harris *Wilmott* 1937 (BM) then south of Loch Laxdale, South Harris *Wilmott* 1939 (BM); North Harris, Glen Meavaig, *Pankhurst* 1984 (BM); North Uist, Loch Obisary, *J.W.Clark*, 1973 (BM); South Uist, Allt Volagir, Beinn Mhor, *G. Taylor* 1951 (BM);   9, 12, 16, 19

*DENNSTAEDTIACEAE*
**Pteridium** Gled. ex Scop.
**P. aquilinum** (L.) Kuhn
Bracken.
Raineach Mhòr.
Native.
Well-drained, often sheltered slopes, not very frequent.
Berneray and Mingulay to Lewis and St Kilda   1–7, 9–13, 15–24, 27, 28

*THELYPTERIDACEAE*
**Phegopteris** (C. Presl) Fée
**P. connectilis** (Michx) Watt
*Thelypteris phegopteris* (L.) Slosson
Beech Fern.
Raineach Fhaidhbhile.*
Native.
Shady mountain rocks.
Barra to Lewis   3–5, 7–9, 12, 16, 19, 21

**Oreopteris** J. Holub
**O. limbosperma** (Bellardi ex All.) J. Holub
*Thelypteris limbosperma* (All.) Fuchs
Lemon-scented Fern.
Crim-raineach, Raineach an Fhàile.
Native.
Moorland, mountains and streamsides.
Barra to Lewis   2, 3, 7, 9, 12, 16, 18, 19, 21

*ASPLENIACEAE*
**Asplenium** L.
**A. scolopendrium** L.
*Phyllitis vulgare* Hill; *P. scolopendrium* (L.) Newman
Hart's-tongue.
Teanga an Fhèidh.
Native.
Sandray to Lewis   3, 5, 8, 9, 11, 12, 16, 18, 19, 21, 23

**A. adiantum-nigrum** L.
Black Spleenwort.
Raineach Uaine.
Native.
Stone embankments, rocks by sea and on mountain rocks.
Pabbay (Mingulay) to Lewis and St Kilda   3, 7, 9, 11, 12, 16, 18, 19, 21, 22, 23, 28

**A. marinum** L.
Sea Spleenwort.
Raineach na Mara.
Native.
Rocks by sea.
Berneray and Mingulay to Lewis and St Kilda   1, 3, 4, 6, 9, 12–14, 16, 18, 19, 21–24, 27, 28

**A. trichomanes** L.
Maidenhair Spleenwort.
Dubh-chasach, Lus a' Chorrain.
Native.
Walls and stone embankments.
The separation of the two subspecies of this taxon relies on the work done by *Lovis* and material not seen by him will probably be the tetraploid subspecies **quadrivalens** D. E. Meyer.

subsp. **trichomanes**
This is much the rarer of the two subspecies, and a calcifuge. Lovis annotated a specimen from Mealasta Island, Uig, Lewis *Wilmott* 1939 (BM) thus:
'*A. trichomanes* L. sensu lato. This specimen is undeterminable but could conceivably be subspecies **trichomanes** which is not

certainly known from the far north-west. This locality is worth refinding for critical study, *Lovis* 1977.

subsp. **quadrivalens** D. E. Meyer
Barra to Lewis   3, 5, 7, 9, 12, 15, 16, 18, 19, 21

**A. viride** Hudson
Green Spleenwort.
Ur-thalmhainn.
Native.
Lower rocks, Tomnaval, Scaladale, North Harris *F. Druce* 1937 (BM)

**A. ruta-muraria** L.
Wall-rue.
Rù Bhallaidh.
Native.
Rocks and stone buildings.
North Uist to Lewis   3, 5, 9, 12, 15–17

*WOODSIACEAE*
**Athyrium** Roth
**A. filix-femina** (L.) Roth
Lady-fern.
Raineach Moire.
Native.
Stream banks, mountain cliffs and sea cliffs.
Barra to Lewis and St Kilda.   1–7, 9, 10, 12, 13, 15, 16, 18, 19–24, 27, 28

**Gymnocarpium** Newman
**G. dryopteris** (L.) Newman
*Thelypteris dryopteris* (L.) Slosson
Oak Fern.
Sgeamh Dharaich.
Native.
Scarp *W.S. Duncan* 1899 (BM); gully on Cracaval, Uig, Lewis *Crabbe* 1939 (CGE); Glen Tealasdale, Uig, Lewis *Wilmott and Campbell* 1939 (CGE); South of Loch Corodale, South Uist, *Braithwaite*, 1983.   3, 8, 19

**Cystopteris** Bernh.
**C. fragilis** (L.) Bernh.
Brittle Bladder-fern.
Frith-raineach.
Native.
Mountain cliffs.
Barra to North Harris and St Kilda   9, 12, 19, 21, 23, 28

*DRYOPTERIDACEAE*
**Polystichum** Roth
**P. aculeatum** (L.) Roth
Hard Shield-fern.

Ibhig Chruaidh.
Native.
Castle Park, Stornoway, Lewis *Campbell* 1938 (BM)
This was the only record accepted in the *Fern Atlas* (Jermy et al., 1978); the literature records from *W. A. Clark and J. W. H. Harrison* from South Uist to North Harris and Taransay were rejected. However there is a specimen, Glen to Loch Trollamarig, North Harris *J. W. H. Harrison* in Herb. *W. A. Clark* (BM).   5, 9.

**Dryopteris** Adanson
**D. filix-mas** (L.) Schott
Male-fern.
Marc-raineach.
Native.
Sea cliffs and stone embankments.
Sandray to Lewis   1, 4, 5, 7, 12, 13, 16–19, 21–23.

**D. affinis** (Lowe) Fraser-Jenkins
*D. pseudomas* sensu Holub & Pouzar pro parte; *D. borreri* auct. angl.
Scaly Male-fern.
Mearlag.
Native.
Sea and mountain cliffs, streamsides and stone embankments.
Distribution of the aggregate species:   3, 8, 9, 12, 19, 21, 23
This complex group of subspecies is only now being elucidated by the work of *Fraser-Jenkins* which this treatment follows. As the identification is so critical the treatment relies on specimens named by *Fraser-Jenkins* only.

subsp. **affinis**
Valtos Glen, Uig, Lewis *Crabbe* 1939 (BM); Sron Scourst, North Harris *Pankhurst* 1984 (BM); Gill Laxdale, South Harris *Pankhurst* 1984 (BM); near Loch Skiport Pier, South Uist *Crabbe* 1947 (BM); Howmore, South Uist *Chater* 1983 (BM)   3, 9, 12, 19

subsp. **borreri** (Newman) Fraser-Jenkins
Valtos Glen, Uig, Lewis *Crabbe* 1939 (BM 2 sheets); Scarp *W. S. Duncan* 1892 (BM) Obbe, South Harris *W. S. Duncan* 1889 (BM)   3, 8, 12

subsp. **cambrensis** Fraser-Jenkins
Tealasdale River, Uig, Lewis *Pankhurst* 1984 (BM); Glen Meavaig, North Harris *Pankhurst* 1984 (BM); Hecla, South Uist *Crabbe* 1947

(BM); Allt Volagir, South Uist *Wilmott, Campbell, Warburg and Crabbe;* near Loch Skiport, South Uist *Chater* 1983 (BM); Allt Heiker, Ersary, Barra *Chater* 1983 (BM); Ard Mhor, Barra *Wilmott* 1938 (BM).   3, 9, 19, 21.

**D. oreades** Fomin
(*D. abbreviata* sensu auct. plur. non DC.)
was recorded from Strone Scourst, North Harris *Campbell* 1938 (BM) det. *Corley* and is represented there by a dot in the BSBI Atlas, but has been redetermined by *Paul* as **D. affinis.**

**D. carthusiana** (Villar) H. P. Fuchs
*D. lanceolatocristata* (Hoffm.) Alston
Narrow Buckler-fern.
Native.
Drinishader, South Harris *M. S. Campbell* 1972 (BM); *Shoolbred* recorded it from Benbecula and North Uist and *J. W. H. Harrison* recorded it from Sandray to Lewis but there are no specimens; Berneray (Smith, 1987).
12, 16, 18, 24.

**D. dilatata** (Hoffm.) A. Gray
*D. austriaca* sensu auct. plur. non Jacq. (1764)
Broad Buckler-fern.
Raineach nan Radan.
Native.
Sea and mountain cliffs, stream banks, rocky moorland and stone embankments.
Barra to Lewis and St Kilda.   1, 3, 5–10, 12, 13, 16–18, 19, 21–24, 28

**D. expansa** (C. Presl) Fraser-Jenkins & Jermy
*D. assimilis* S. Walker
Northern Buckler-fern.
Native.
Coastal rocks to Mountains.
Barra to North Harris and St Kilda   9, 12, 16, 18, 19, 21, 28

**D. aemula** (Aiton) O. Kuntze
Hay-scented Buckler-fern.
Raineach Phreasach.*
Native.
Stream banks, moorland and sea cliffs.
South Uist to Scarp, Lewis and St Kilda   1, 3, 5, 8, 10, 12, 15–19, 21, 28

*BLECHNACEAE*
**Blechnum** L.
**B. spicant** (L.) Roth
Hard Fern.
Raineach Chruaidh.

Native.
Rocky moorland and streamsides.
Berneray and Mingulay to Lewis and St Kilda   1, 3–7, 9–13, 15–24, 27, 28

**GYMNOSPERMAE**
*CONIFEROPSIDA*
*PINACEAE*
**Larix** Miller
**L. decidua** Miller
Larch.
Learag.
Introduction.
Planted and naturalised on coastal rocks.
Stornoway, Lewis, Cunningham (1978) and *Pankhurst* 1984 (BM)   5

**L. × eurolepis** A. Henry
(*L. decidua* Miller × *kaempferi* (Lamb.) Carrière)
Larch.
Planted.
Head of Bagh Huilavagh, Barra *Chater* 1983 (BM)   21

**Pinus** L.
**P. sylvestris** L.
Scots Pine.
Giuthas.
Plantations, and possibly native.
Only one herbarium specimen seen, from Lochskiport, South Uist *Cannon & Cannon* 1983 (BM). Also reported as stunted growth on island in S. W. corner of Loch Laxavat Ard, Lewis, where it might be native (Biagi et al. 1985); Lewis, Stornoway Castle, Cunningham (1978).   2, 5, 12, 19, 21, 22

**P. nigra** Arnold
subsp. **laricio** (Poiret) Maire
Corsican Pine.
Planted.
Lewis, Stornoway Castle, Cunningham (1978), at the head of Little Loch Roag, Blake (1966); Barra, Ard Mhor plantation, *Wilmott* 1938 (BM)   3, 5, 21

**P. mugo** Turra
Mountain Pine.
Planted.
Lewis, Stornoway Castle, Cunningham (1978); South Harris, plantation by Lodge, Borve, *Campbell* 1938 (BM),   5, 12

*CUPRESSACEAE*
**Juniperus** L.
**J. communis** L.

subsp. **communis**
Juniper.
Aiteann.
Recorded in the Critical Supplement to the Atlas of the British Flora from the following zones, but most records seem to be the next subspecies.
South Uist, North Harris and Uig, Lewis   3, 9, 19

subsp. **alpina** (Sm.) Čelak
subsp. *nana* Syme; *J. nana* Willd.; *J. sibirica* Burgsdorf
Juniper.
Iubhar Beinne.
Native.
Mingulay to Lewis.   1, 3–5, 7–10, 12, 13, 16, 18–24

**MAGNOLIOPHYTA**
*RANALES*
RANUNCULACEAE
**Caltha** L.
**C. palustris** L.
Marsh-marigold.
Lus Buidhe Bealltainn.
Native.
Marshes and lochsides.
Plants have been recorded as *C. radicans, C. palustris* var. *radicans* and var. *guerangerii* Bor. Woodell and Kootin-Sanwu (1971) place the first two in **C. palustris** subsp. **minor** (Miller) Clapham and the last in **C. palustris** subsp. **palustris**. However, their view is that neither subspecies should be maintained and that the procumbent form which roots at the nodes should be called *C. palustris* var. *radicans* (Forst.) Beck and we concur with this view.
On all the major islands.   1–4, 6, 7, 9–16, 18–22, 24, 27, 29.

**Anemone** L.
**A. nemorosa** L.
Wood Anemone.
Flùr na Gaoithe.
Probably introduced.
Recorded in the BSBI Atlas for the Stornoway area from a field record in about 1950. This record is highly doubtful, and requires confirmation.

**Ranunculus** L.
**R. acris** L.
Meadow Buttercup.
Buidheag an t-Samhraidh.
Native.

Dunes, machair, damp grassland and in mountain grassland.
Most records are of subsp. **acris**. There are a few specimens which are more densely hairy, but the significance of this is probably little more than exposure.   1–7, 9, 10, 12, 13, 15–24, 27, 28.

**R. repens** L.
Creeping Buttercup.
Buidheag.
Native.
Hedges, waste ground, 'lazy beds' and machair   2–13, 15–25, 27–29.

**R. bulbosus** L.
Bulbous Buttercup.
Fuile-thalmhainn.
Native.
Seems restricted to the machair and sub-maritime sites.
var. **dunensis** Druce is recorded from North Harris, Scarp *W.S.Duncan* 1891 det. *Austen* (BM). As the species flowers early in the Islands it is probably under-recorded.
Barra, Mingulay, the Uists, Harris and Uig.   3, 4, 8, 9, 12, 16, 19, 21, 24, 29

**R. sardous** Crantz
Hairy Buttercup.
Native.
Rare, one record only; Barra, in rough pasture, *Bush* 1949 (OXF).   21

**R. flammula** L.
Lesser Spearwort.
Glaisleun.
Native.
Wet places; loch and stream sides, flushes and marshes.
The records are mainly for subsp. **flammula**. Subsp. **scoticus** (E. S. Marshall) Clapham is recorded from North Uist by *Shoolbred* (BM) and from Uig by *Bangerter* (BM), and subsp. **minimus** (A. Benn) P. A. Padmore from Butt of Lewis, *Campbell*, (BM) and from North Uist by *Shoolbred* (BM) and *Murray* (E). There are also records for vars. **pseudoreptans** (Lewis) and **petiolaris** (Harris). See Padmore (1957).   1–10, 12–24, 27–29.

R. reptans L.
There is a literature record from a lochan in Scaladale, North Harris (Harrison, 1954b). Possibly an error based on a form of the last species. No herbarium specimen has been

seen, and confirmation of this species is required.

**R. sceleratus** L.
Celery-leaved Buttercup.
Torachas Biadhain.
Native.
Wet mud and loch margins.
Lewis, Port Sto, *Balfour* and *Babington*, 1841; South Uist *Somerville* July 1888 (BM), Na-Baighe-Dubha, Loch Eynort, *Warburg* 485 (OXF).
Benbecula and the Uists and Lewis.    1, 16, 18, 19, 24

**R. hederaceus** L.
Ivy-leaved Crowfoot.
Fleann Uisge Eidheannach.
Native.
Ditches and muddy places.
South Uist, Barra and Lewis.    1, 6, 7, 19, 21, 24.

**R. trichophyllus** Chaix subsp. **trichophyllus**
Thread-leaved Water-crowfoot.
Lìon na h-Aibhne.
Native.
Coastal streams, mesotrophic lochs.
Loch St. Clair, Barra *Campbell* 1978 (BM); Loch Ardvule, South Uist *Warburg* 141 (OXF).
Barra and South Uist.    3, 11, 14–16, 18–22, 24, 29.

**R. trichophyllus** subsp. **drouetii** (Godron) Clapham
Native.
Ponds in dunes.
Barra, the Uists, Berneray, South Harris and Lewis.    1, 2, 15, 16, 19, 21

**R. aquatilis** L.
(incl. *R. peltatus* Schrank.)
Common Water-crowfoot.
Fleann Uisge.
Native.
Mesotrophic lochs by the sea and on machair.
South Uist, Loch an Rubha Ardvole *Chater* (BM); Barra, Traigh Scurrival *Campbell* (as *R. peltatus*, (BM)); Vatersay (*Conacher*, 1980).
Barra, Uists and Vatersay.    16, 19, 21, 22, 24.

**R. baudotii** Godron
Brackish Water-crowfoot.
Native.
Near the sea in brackish water.
Benbecula and the Uists, with one record from Coll, Lewis.    16, 18, 19

**R. ficaria** L.
Lesser Celandine.
Searragaich, Gràn Aigein.
Native.
Roadside and streamside banks.
Widespread to Hirta.    1, 3–5, 8, 9, 11–16, 18–22, 24, 28, 29.
subsp. **bulbifer** (Marsden-Jones) Lawalrée is represented by a specimen from South Harris near Tarbert *Marquand* 1938 (BM).

**Thalictrum** L.
**T. alpinum** L.
Alpine Meadow-rue.
Rù Ailpeach.
Native.
Mountains, damp shady ledges and flushes descending to sea level.
South Harris, Roneval *Campbell* near summit (BM); Scarista *W.S.Duncan* rocks at sea level (BM); North Harris, Geoan Dubh, *Warburg* 657 (OXF).    5, 9, 12, 16, 19.

**T. Minus** L.
Lesser Meadow-rue.
Rù Beag.
Native.
Machair and sand dunes.
Widespread through the islands.
Material from Barra, Lewis and South Harris has been separated as subsp. **arenarium** (Butcher) Clapham (including var. *dunense* auct.); this is said to differ from subsp. **minus** (including *T. montanum* Wallr.) in the panicle branching from below the middle of the stem but the distinctions are not clear cut.    1–3, 9, 11–16, 18–24, 29.

*NYMPHAEACEAE*
**Nymphaea** L.
**N. alba** L.
White Water-lily.
Duilleag-bhàite Bhàn.
Native.
Oligotrophic and mesotrophic lochs.
From Lewis to Barra.
Material from Lewis has been separated as *N. occidentalis*(Ostenf.) Moss but this is now considered as part of the variation of **N. alba**.    1–5, 7, 10–12, 16–21.

**Nuphar** Sm.
**N. lutea** (L.) Sm.
Yellow Water-lily.
Duilleag-bhàite Bhuidhe.
Native.
North Uist; Loch Fada *MacCuish & Campbell*

(BM); Loch Maddy *Shoolbred* (BM); lochan North side of A867 *Chorley* (BM): South Uist *Campbell* (CGE).

Perhaps from more mesotrophic lochs than the last species.

North and South Uist. 16, 19.

RHOEADALES
*PAPAVERACEAE*
**Papaver** L.
**P. rhoeas** L.
Common Poppy.
Meilbheag.
Probable introduction.
Scarce. Recorded in the BSBI Atlas for Eriskay and North Uist from field records. These might have been errors for P. dubium. There are no herbarium specimens, and confirmation is required.

**P. dubium** L.
Long-headed Poppy.
Crom-lus Fad-cheannach.
Colonist.
Arable fields.
Throughout the islands. 1–3, 9, 11, 12, 15, 16, 18, 19, 21, 22, 29.

**P. argemone** L.
Prickly Poppy.
Crom-lus Calgach.
Probably a colonist.
North Uist; Oronsay Island *Babington* 1841 (CGE), Trumisgarry (Balfour and Babington, 1844).
No modern records. 16.

**Meconopsis** Vig.
**M. cambrica** (L.) Vig.
Welsh Poppy
Crom-lus Cuimreach.
Introduced.
Lewis, Stornoway Castle woods, Cunningham (1978). 5

*FUMARIACEAE*
**Fumaria** L.
**F. bastardii** Boreau
Tall Ramping-fumitory
Fuaim an t-Siorraimh.
Colonist.
Arable fields.
South Uist to Lewis. 1–4, 8, 9, 11, 12, 14–16, 18, 19, 21

**F. muralis** Sonder ex Koch subsp. **boraei** (Jordan) Pugsley

Common Ramping-fumitory.
Dearag Thalmhainn.
Colonist
North Uist, field records only; Barra, Traigh Scurrival *M. S. Campbell* (BM); Vatersay *M. S. Campbell* (BM);
There are several ill-localised literature records by J. W. H. Harrison (1941a). 16, 18, 21, 22, 24.

**F. officinalis** L.
Common Fumitory.
Lus Deathach-thalmhainn.
Colonist.
Cultivated land and roadsides.
Lewis, Meavaig, *Wilmott and Warburg* 80 (OXF); Barra *Somerville* (BM); causeway between Benbecula and North Uist *Duncan* (BM). The subsp. **wirtgenii** (Koch) Arcangeli is recorded. 3, 16–21, 24

**F. parviflora** Lam.
Fine-leaved Fumitory.
Recorded at Seilebost in South Harris in 1941 (Harrison, 1941a); very doubtful and not supported by a specimen. 12?

*CRUCIFERAE*
**Brassica** L.
**B. napus** L.
Rape.
Raib, Snèap Suaineach.
Denizen.
Benbecula and North Uist (*Shoolbred*, 1895); South Uist, Stilligarry *Davis* (E); Barra: *Somerville* (BM), *Davis* (E). 12, 16, 18–22

**B. rapa** L.
Wild Turnip.
Snèap Fiadhain.
Denizen
Arable fields on machair.
Barra, Benbecula, the Uists and South Harris, also literature records for Lewis and Berneray (Harrison, 1941a); 12, 15, 16, 18, 19, 21, 28.

**Sinapis** L.
**S. arvensis** L.
Charlock.
Sgeallan.
Colonist.
There are no records of var. **orientalis**, which lacks hairs on the mature fruit.
Lewis, Harris, Benbecula. 1, 3, 6, 12, 15, 16, 18, 19, 20, 21, 24, 29.

**S. alba** L.
Sgeallan Bàn.

White Mustard.
Denizen.
Literature records only. Butt of Lewis (Balfour and Babington, 1844); Benbecula 'an escape' (*Shoolbred*, 1895).   1, 18.

**Raphanus** L.
**R. raphanistrum** L.
Wild Radish.
Meacan Ruadh Fiadhain.
Possibly native
Arable fields.
Barra, Uist, Benbecula, Harris and Lewis. 2–4, 6, 7, 9, 10, 12, 15, 16, 18, 21, 22, 29.

**R. maritimus** Sm.
Sea Radish.
Meacan Raguim Uisge.
Native.
Rare.
South Harris, Northton, *Duncan* (BM, CGE) 1892; South Harris, Harrison et al. (1941a) record it from 'Seaward side of Luskentyre dunes'; Ensay, *Morton*, 1958; Eriskay, Prince's Bay, *Webster*.   12, 14, 20.

**Crambe** L.
**C. maritima** L.
Sea-kale.
Càl na Mara.
Native.
Foreshore.
Only known from North Uist; Eilean a' Mhorain and Loch Maddy
Eilean a' Mhorain *Nicholson* and *Campbell* (BM); Lochmaddy *Burlingham* (Campbell, 1937b).   16.

**Cakile** Miller
**C. maritima** Scop.
Sea Rocket.
Fearsaideag.
Native
Sandy driftlines
Common through the islands.
Some specimens from throughout the range have been separated as subsp. **integrifolia** (Hornem.) Hyland (*C. edentula* auct.), see Allen (1952). The taxonomic situation is unclear.   1, 3, 9, 11, 12, 14–16, 18–22, 24, ?28, 29.

**Lepidium** L.
**L. campestre** (L.) R. Br.
Field Pepperwort.
Lus a' Phiobair.
Doubtfully native.

The only record is from Castlebay, Barra *Wilmott* 380718A (BM); Also confirmed J. M. Mullin.   21.

**Coronopus** Haller
**C. squamatus** (Forskål) Ascherson
Swine-cress.
Muic-bhiolair.*
Colonist.
Barra: Castle Bay *Campbell* (BM); North Bay *Edinb. Univ. Biol. Soc.* (E); South Harris, Horgabost, Harrison (1957).   12, 21.

**C. didymus** (L.) Sm.
Lesser Swine-cress.
Muic-bhiolair as Lugha.*
Colonist.
Barra: Castlebay, first record in 1978, Conacher (1980), also from road verge, Castle Bay *Chater* 1983 (BM).   21.

**Thlaspi** L.
**T. arvense** L.
Field Penny-cress.
Praiseach Fèidh.
Probably introduced.
Rare, one record only; South Harris, Tarbert, field record *Perring*, 1959.   12

**Capsella** Medicus
**C. bursa-pastoris** (L.) Medicus
Shepherd's-purse.
An Sporan, Lus na Fala.
Native.
Arable fields and waste places.
Common from Uig southwards.   2–4, 6, 7, 9, 10, 13–16, 18–22, 24, 27, 28

**Cochlearia** L.
This genus remains in a taxonomically confused state. See Gill (1971) and Gill et al. (1978).
**C. officinalis** L.
subsp. **officinalis**
Common Scurvygrass.
Maraiche, Carran.
Native.
Coastal rocks, mountain rock ledges and mortared walls.
Common through the islands.   3–5, 7, 9, 12–25, 28, 29.

subsp. **alpina** (Bab.) Hooker
(*C. alpina* (Bab.) H. C. Wats., non *C. pyrenaica* DC.)
Alpine Scurvygrass.
Native.

Most of the high level material of Cochlearia seems to be the former subspecies but a specimen from St. Kilda, *Barrington* (BM), may represent this taxon.    ?28

**C. scotica** Druce (*C. groenlandica* auct.)
Scottish Scurvygrass.
Native.
Salt-marshes, coastal turf and rock ledges.
Fairly common.    1, 3, 4, 6–10, 12–16, 18, 19, 24, 25

**C. danica** L.
Danish Scurvygrass.
Native.
Coastal.
Berneray to Harris.    4, 12, 13, 15, 16, 18, 19, 21, 22, 24, 28

**C. anglica** L.
English Scurvygrass.
Native.
South Uist to Lewis.    3, 10, 12, 15, 16, 19, 20, 24.

**C. atlantica** Pobed.
Atlantic Scurvygrass.
Native. South Uist, Pollachar *Campbell* (BM) and Lewis, Uig, Crowlista *Campbell* (BM), both det. Pobedimova. Has sessile and amplexicaul stem leaves, white spathulate petals, broadly ellipsoid to subglobose fruit and seeds c. 1mm. It has been suggested that this species is a mixture of C. scotica and C. officinalis subsp. officinalis, but further investigation is needed.    3, 19.

**Subularia** L.
**S. aquatica** L.
Awlwort.
Ruideag, Lus a' Mhinidh.*
Native.
Margins of lochs.
May be more frequent than the records suggest, as it is not easy to find. Mountain range, North Harris *Duncan* (BM); Lewis, Loch Urrahag *Fox* (BON). Also published records from South Uist, Stoneybridge, (Harrison, 1941a), Lewis (Perring, 1961; Biagi et al. 1985) and MacGillivray (1830). 1–3, 9, 12, 19

**Draba** L.
**D. incana** L.
Hoary Whitlowgrass.
Native.
Mountain rocks and dunes.

South Uist, Ben More, rocks in corrie *Warburg* 664 (BM, OXF), Berneray, on the dunes *J. W. H. Harrison* (K). Macgillivray (1830) records it from Ben Capval, Harris and Harrison et al. (1939) from Fuday. The most recent record is from Prince's Strand, Eriskay, by *Barron* in 1977.    12, 15, 19, 20, 21

**Erophila** DC.
**E. verna** (L.) Chevall.
Common Whitlowgrass
Biolradh Gruagain.
Native.
According to Rich and Rich (1988), the subspp. **verna**, **praecox** (Stev.) DC. and **spathulata** (Láng) Walters are best ignored and treated as part of the variation of a single species. The species is probably much under-recorded as it flowers so early.
North Uist, north side of Loch Sandary *McDonald and Brockie* (BM); literature records exist for Monach Islands and Harris (Clark, 1939c) and on dunes at Daliburgh, South Uist (Harrison, Harrison, Clark and Cooke, 1942b). There is also a literature record for subsp. **spathulata**; North Uist, Sollas (Harrison, 1941a).
Records for subsp. **praecox**; North Uist, between Knockintorran and Paible *Fergusson and Brockie* (BM), South Harris, Rodel, wall top *Campbell* (BM), Lewis, North Tolsta, Traigh Mhor *Campbell* (BM).    1, 12, 16, 19, 29

**E. glabrescens** Jordan
(**E. quadriplex** Winge)
Whitlowgrass.
Biolradh Gruagain.
Native.
Wall tops and beaches.
Barra; Greian Head, *Edinburgh Univ. Biol. Soc.*, 1938 (E); Lewis, North Tolsta, *M. S. Campbell*, 1939 (BM); both det. *Elkington*.
Probably under-recorded.    1, 21

Records for aggregate of all spp.    1, 3, 12, 16, 19, 29.

**Cardamine** L.
**C. pratensis** L.
Cuckooflower.
Flùr na Cuthaig.
Native.
Marshes, flushes and streamsides.
Barra, North Bay *Edinburgh Univ. Biol. Soc.* (E), South Uist, Loch Ceann a' Bhaigh *Wilmott, Campbell, Warburg and Crabbe* (BM),

Benbecula, near Creagory *Vincent* (BM). Some BM specimens have been determined by *Allen* as hybrids of **C. pratensis** and **C. nymanii** Gand. (**C. polemoniodes** Rouy). These specimens all seem to be part of the normal variation of **C. pratensis**, D. E. Allen in a personal comm. stated that in North Britain the species is a complex of hybrids between different ploidy levels and further work is required to tie the morpho-metric groups into the cytospecies. Harrison (1941a) states 'all islands except Berneray.' 1–7, 9, 10, 12, 13, 15–24, 29

**C. flexuosa** With.
Wavy Bitter-cress.
Searbh-bhiolair Chasta.**
Native.
Mountain rock ledges, parks.
Lewis, Uig, Harris, Uists, Benbecula and Barra.    1, 3, 5, 7, 9, 12, 16, 18, 19, 21, 23, 29.

**C. hirsuta** L.
Hairy Bitter-cress.
Searbh-bhiolair Ghiobach.**
Native.
Benbecula, Uists, Harris and Lewis.    1, 5, 9, 11, 12, 15, 16, 18, 19, 21, 28, 29.

**Barbarea** R. Br.
**B. vulgaris** R. Br.
Winter-cress.
Treabhach.
Casual.
Field record only. Balvanich, Benbecula, roadside verge, *Pankhurst* 1979.    18.

**B. verna** (Mill.) Aschers.
American Winter-cress.
Treabhach.
Colonist.
Barra, Castlebay *M. S. Campbell* 1978 (BM) Only record.    21.

**Cardaminopsis** (C. A. Meyer) Hayek
**C. petraea** (L.) Hiit.
Northern Rock-cress.
Biolair na Creige Thuathach.
Native.
Only records are from literature.
Harris, Mullach an Langa (Balfour and Babington, 1844), Clisham *Campbell* in Perring (1961), summit cliffs of Uisgnaval More (Harrison, Harrison and Clark, 1941a).    9.

**Arabis** L.
**A. hirsuta** (L.) Scop.

Hairy Rock-cress.
Biolair na Creige Ghiobach.
Native.
Machair and dunes.
Widespread on machair from Barra to North Uist and with a literature record from Lewis (Clark and Harrison, 1940).    3, 4, 11, 12, 15, 16, 21.

**Rorippa** Scop.
**R. sylvestris** (L.) Besser
Creeping Yellow-cress.
Biolair Bhuidhe Ealaidheach.
Native?
Streamsides, waste places.
South Harris, Borve *J. W. H. Harrison* (K) 1950; Borve, sandy bank by bridge over Borvebeg Burn *Chater & Pankhurst* (BM) 1980. Also published for Tarbert, North Harris, (Harrison, 1956) and still there, *Pankhurst*, 1984 (BM), outside the police station! Both det. *Jonsell*, it is interesting to see these records separated by 30 years; Ensay, *Morton*, 1958.    9, 12, 14.

**R. islandica** (Oeder ex Murray) Borbás
Northern Yellow-cress.
Native.
Monach Islands *Randall* 1969 and 1970 det. *Jonsell* (Randall, 1974).    29.

**Nasturtium** R. Br.
**N. officinale** R. Br.
*Rorippa nasturtium-aquaticum* (L.) Hayek
Water-cress.
Biolair.
Native.
Ditches.
Barra to Harris.    3, 6, 11, 12–16, 18–21, 23, 24, 29.

**N. microphyllum** (Boenn.) Reichenb.
*R. microphylla* (Boenn.) Hyl.
Narrow-fruited Water-cress.
Native.
Roadside ditches.
Barra to Lewis.    6, 16, 18, 19, 21

**N. officinale × microphyllum**
*R. × sterilis* Airy-Shaw
Hybrid Water-cress.
Native.
Eriskay, Ditch near burial ground *Cannon and Cannon* (BM); SW Barra *Davis* (E); Benbecula, Pennylodden *Wilmott et al.* (BM) Probably under-recorded.    8, 18, 20, 21.

**Hesperis** L.
**H. matronalis** L.
Dame's-violet.
Garden escape.
Barra; field record BSBI meeting 1978; North Uist, Lochmaddy, *J. W. Clark* and *Murray*, 1982, BM.   16, 21

**Sisymbrium** L.
**S. officinale** (L.) Scop.
Hedge Mustard.
Meilise.
Native.
Waste places.
Barra to Lewis.
Some *Wilmott* collections have been determined as var. *leiocarpum* DC. with glabrous siliquae valves notably: Barra, Castlebay, 1938 and Berneray, 1939.   3, 6, 12, 15–17, 19, 21, 22, 29

**Arabidopsis** (DC.) Heynh.
**A. thaliana** (L.) Heynh.
Thale Cress.
Biolair Thailianach.
Native.
Literature records from South Uist and South Harris (Harrison, 1956).   12, 19.

VIOLALES
*VIOLACEAE*
**Viola** L.
V. rupestris Schmidt
Teesdale Violet.
Recorded by *Druce* from Harris, Husinish (Druce, 1929); thought to be an error for **V. riviniana** which, in its slightly pubescent form (f. *villosa* N. W. & M.), Druce equated with *V. arenaria* and hence this species.

**V. riviniana** Reichenb.
Common Dog-violet.
Dail-chuach.
Native.
Common throughout the Islands.
Specimens collected in Lewis, Uig by *Crabbe* were determined as forma **villosa** Neuman; South Harris, Roneval *Campbell* as subsp. **minor** (Gregory) Valentine by Valentine.
1–9, 11–24, 28, 29.

**V. reichenbachiana** Jordan ex Boreau
Early Dog-violet.
Native.
Only known in literature; records from South Uist, near Ormiclate Castle (Harrison, Harrison, Cooke and Clark, 1939); Loch Eynort (Harrison, 1941a); Harris, East Loch Tarbert (ibid.).
Seems likely to be in error.

**V. canina** L.
Heath Dog-violet.
Sàil-chuach.
Native.
Dunes and sandhills (one site on a wall).
Mingulay to Lewis.   3, 8, 16, 18, 21, 24

**V. palustris** L.
Marsh Violet.
Dail-chuach Lèana.
Native.
Marshes, flushes and damp grassland.
Barra to Lewis and St Kilda.   1–7, 9, 11–13, 15–21, 24, 28, 29.

**V. lutea** Huds.
Mountain Pansy.
Native.
Recorded from North Uist and Benbecula in the last century, but at least some of the BM specimens have been redetermined as **Viola tricolor** L. subsp. **curtisii** (E. Forster) Syme and as the records are from sand-dunes the rest probably belong there too.

**V. tricolor** L. subsp. **tricolor**
Wild Pansy.
Goirmean-searradh.
Native.
Machair and arable fields.
Barra to Lewis.   1, 5, 11, 16, 19–21.

**V. tricolor** L. subsp. **curtisii** (E. Forster) Syme
Goirmean Buidhe.
Native.
Machair
Common on machair from Barra to Lewis.
1, 11–13, 15, 16, 18–21, 29.

**V. arvensis** Murray
Field Pansy.
Luibh Chridhe.
Native (or colonist in arable fields)
Arable fields and gardens.
Vatersay and Barra to Lewis.   2, 12, 15, 16, 18, 19, 21, 22

POLYGALALES
*POLYGALACEAE*
**Polygala** L.
**P. vulgaris** L. (inc. **P. oxyptera** Reichenb.)
Common Milkwort.

Lus a' Bhainne.
Native.
Grassy and often more base-rich situations.
Common from Barra to Lewis and on St
Kilda.    1–6, 9, 11–13, 15, 16, 19–24, 28, 29.

**P. serpyllifolia** Hose
Heath Milkwort.
Siabann nam Ban-sìdh.
Native.
Moorland, marshy grassland and short turf.
Very common from Barra to Lewis and St
Kilda.
Probably more common than last species.
1–7, 9–12, 15–24, 28, 29.

CISTIFLORAE
*GUTTIFERAE*
**Hypericum** L.
**H. perforatum** L.
Perforate St. John's-wort.
One record only, seacliffs south of Stornoway,
Harrison (1957).
Requires confirmation.

**H. pulchrum** L.
Slender St. John's-wort.
Lus Chaluim Chille.
Native.
Streambanks, rock ledges and gardens.
Common from Mingulay to Lewis.
Extreme habitat forms with prostrate habit
and reduced 3-flowered cymes recorded
from St. Kilda *Gladstone* (K) and from the
summit of Suainaval, Lewis (BM) have
been separated as var. **procumbens** Rostrup
(Beeby, 1888) but according to Robson (pers.
comm.) are so connected by intermediates as
to not be worthy of separation even as a
forma.    3–5, 7, 9–13, 16–19, 21–24, 27, 28.

**H. elodes** L.
Marsh St. John's-wort.
Meas an Tuirc-allta.
Native.
Boggy pools.
Vatersay to North Uist with a literature
record from Harris *Campbell* (Perring, 1961).
12, 16, 19, 21, 22.

CENTROSPERMAE
*ELATINACEAE*
**Elatine** L.
**E. hexandra** (Lapierre) DC.
Six-stamened Waterwort.
Native first record *W. S. Duncan* 1895
On muddy bottoms of lochs.

Barra, Lochan nam Faoilleann *Sinclair* (E),
Loch St. Clair *Edin. Univ. Biol. Soc.* (E).
South Uist; Loch Torornish and L. Altabrug,
Chamberlain et al. (1984) and literature
record, Loch Ollay (Harrison, Harrison,
Cooke and Clark, 1939); North Harris, small
loch in Glen Laxdale Duncan (1896).
Rare but may well be under-recorded.    12,
19, 21.

*CARYOPHYLLACEAE*
**Silene** L.
**S. dioica** (L.) Clairv.
Red Campion.
Cìrean Coilich.
Native.
Shiant Islands *Babington* (BM), Lewis *Currie*
(1981a), South Harris *Shoolbred* (1895) and
South Uist (Harrison, 1941a).    1, 12, 19, 27.

**S. latifolia** Poiret
*S. alba* (Miller) E. H. L. Krause non Muhl. ex
Britton, *S. pratensis* (Rafn) Godron & Gren.
subsp. **alba** (Miller) Greuter & Burdet
White Campion.
Coirean Bàn.
Native.
From Vatersay to Lewis in arable fields.    1,
12, 16, 18, 19, 21, 22, 24.

**S. vulgaris** (Moench) Garcke subsp. **mari-
tima** (With.) A.& D.Löve
*S. maritima* With.
Sea Campion.
Coirean na Mara, Oigh na Mara.
Native.
On the shore and cliffs, both coastal and on
mountains.
Frequent through the Islands.
From Barra Head on Bernera to Butt of Lewis
and Hirta.    1–4, 6, 9, 11–13, 16, 19, 21–24,
27–29.

**S. acaulis** (L.) Jacq.
Moss Campion.
Coirean Ailpeach.
Native.
Rock ledges; down to sea-level on the west
coast of Lewis.
On Mingulay and Barra and on Lewis and
North Harris.    3, 4, 9, 24, 28.

**Lychnis** L.
**L. flos-cuculi** L.
Ragged-Robin.
Sìoda-lus, Caorag Lèana.
Native.

Marshes and flushes.
Barra to Lewis.
A white form has been reported from South Uist *Somerville* (E). 2–4, 6, 8, 9, 12–24, 27–29.

**Agrostemma** L.
**A. githago** L.
Corncockle.
Lus Loibheach.
Probably always a colonist.
Only record North Uist, Hougharry Bay (*Shoolbred*, 1898). 16.

**Cerastium** L.
**C. arcticum** Lange
Arctic Mouse-ear.
Benbecula, low elevation on Rueval (Harrison, 1941a). Comment by Wilmott on the BM copy – 'I don't believe it, impossible habitat for this species!'

**C. fontanum** Baumg.
*C. vulgatum* auct.; *C. holosteoides* Fries
Common Mouse-ear.
Cluas Luch.
Native.
Grasslands, machairs, roadsides, moorland and rock ledges.
Common through the Islands. 1–10, 12–25, 27, 28.

**C. glomeratum** Thuill.
Sticky Mouse-ear.
Cluas Luch Fhàireagach.
Native.
Mingulay to Lewis.
Not as abundant as the last species. 1, 3–7, 10, 12, 15–22, 24, 29

**C. diffusum** Pers.
*C. atrovirens* Bab.; *C. tetrandrum* Curtis
Sea Mouse-ear.
Cluas Luch Mara.
Native.
Machair, sand dunes and coastal rocks.
Common.
A specimen from South Harris, Luskentyre, on dunes by *Wilmott* 1938 was described by Wilmott (1941) as the type of **C. tetrandrum** var. **pusillum** Wilmott (BM); this appears to be merely a very small flowered form. Another from South Uist, Daleburgh, on dunes *Blackburn* (K) was determined by her as possibly **tetrandrum** × **viscosum** i.e. **diffusum** × **glomeratum** . 1–4, 9, 11, 12, 14–24, 25, 27, 28.

**C. semidecandrum** L.
Little Mouse-ear.
Cluas Luch Bheag.
The only record is Benbecula, machair (Harrison, 1941a); this requires confirmation.

**Stellaria** L.
**S. media** (L.) Vill.
Common Chickweed.
Fliodh.
Native.
Waste places, arable fields and scrub.
Barra to Lewis and Hirta. 3–25, 27–29.

**S. neglecta** Weihe
Greater Chickweed.
Native.
Lewis, Valtos, by pier *Pankhurst* and *Chater* (BM); Mingulay, gullies of west cliffs (*W. A. Clark*, 1938); Berneray, gullies of west cliffs (ibid.); South Uist (Harrison, 1941a). 3, 19, 24.

**S. graminea** L.
Lesser Stitchwort.
Tursarain.
Native.
Lewis, Creed River, Harrison (1954b); North Harris, Amhuinnsuidhe *M. S. Campbell* and *Wilmott* (BM); Barra (*Conacher*, 1980). 5, 9, 21.

**S. alsine** Grimm
Bog Stitchwort.
Flige.
Native.
Streamsides and springs.
From Barra to Lewis and Hirta. 1–7, 9, 10, 12, 14–19, 21, 22, 24, 27, 28.

**Sagina** L.
**S. apetala** Ard. subsp. **apetala**
*S. ciliata* Fries
Ciliate Pearlwort.
Lus Pèarlach.*
Native.
Heathy places.
Only recorded by Harrison et al. (1941a) from North Harris, Laxadale Burn. 9.

**S. apetala** Ard. subsp. **erecta** (Hornem.) F. Hermann
*S. apetala* auct.
Annual Pearlwort.
Lus Pèarlach Bliadhnail.*
Native.

Barra, *Somerville* 9627, 1897 (BM) is the only specimen seen. Literature records are; North Uist; Harris and Taransay *Shoolbred* (1895) and South Uist, on a path (Harrison, 1941a). 11, 16, 19, 21.

**S. procumbens** L.
Procumbent Pearlwort.
Lus Pèarlach Làir.*
Native.
Waste ground, roadsides, rocks and machair. Common everywhere.
The variety **S. procumbens** var. **flexilis** Nolte has been recorded from Barra; *Edin. Univ. Biol. Soc.* BM, Loch na Doirlinn *Wilmott* BM; Benbecula *M. S. Campbell* BM during 1936–8 period. 1–25, 27, 28.

**S. maritima** G. Don fil.
Sea Pearlwort.
Lus Pèarlach Mara.*
Native.
Coastal rocks.
Barra to Lewis and possibly North Rona. 1–4, 6, 8, 10, 12, 16, 18, 19, 21, 24, 25, 29.

**S. subulata** (Sw.) C. Presl
Heath Pearlwort.
Lus Pèarlach Mòintich.*
Native.
Plants which are glabrous or nearly so, instead of being glandular hairy on the sepals, pedicels and leaves, occur frequently. These are known to be commoner in the north of Britain (Harrold, 1978). Walters (1988) points out that a yellow variant, known as cultivar 'Aurea', is currently popular for patio gardens.
From summits down to the shore.    2, 3, 7, 9, 12, 15, 16, 18, 21, 22, 24, 27, 28.

**S. nodosa** (L.) Fenzl
Knotted Pearlwort.
Lus Pèarlach Snaimte.*
Native.
Machair, coastal rocks and marshy grassland. Barra to Lewis.    5, 9, 12, 14, 15, 16, 18–21, 24, 29.

**Minuartia** L.
M. sedoides (L.) Hiern
*Cherleria sedoides* L.
Cyphel.
Listed in Currie (1979) but the only evidence is a note in Miss Campbell's papers 'BEC and Druce 1928'. We have not been able to find any reference or records in the BEC

Report for 1928 nor elsewhere, and can only suggest that this is a copying error.

**Honkenya** Ehrh.
**Honkenya peploides** (L.) Ehrh.
Sea Sandwort.
Lus a' Ghoill.
Native.
Shingle, sand, salt marsh and arable fields. Barra to Harris.    1–3, 4, 6, 11, 12, 15, 16, 18–22, 24, 28, 29.

**Arenaria** L.
**A. serpyllifolia** L.
Thyme-leaved Sandwort.
Lus nan Naoi Alt.
Native.
Waste places, machair and arable fields. Mingulay to Lewis
A specimen from South Uist, Daliburgh, on path to rifle range, 1964 *Crompton* (CGE) has been determined as subsp. **serpyllifolia** by *Sell*.    1, 3, 4, 11, 12, 14–16, 18–22, 24, 29.

**Spergula** L.
**S. arvensis** L.
Corn Spurrey.
Cluain-lìn, Corran-lìn.
Colonist.
Arable fields and gardens.
Mingulay to Lewis and Hirta.    2–7, 9, 10, 12, 15–21, 24, 28, 29.

**Spergularia** (Pers.) J. & C. Presl
**S. rubra** (L.) J. & C. Presl
Sand Spurrey
Corran Gainmhich.
Tracks and roadsides.
Possibly introduced.
There is a specimen from Lewis, near Stornaway, 1950 *J. W. H. Harrison* (K) which appears to be correct; Campbell (1944) recorded it from Scalpay in 1939. Field records from Barra, BSBI meeting 1978; Eye peninsula, 1959 and from Loch Mor an Iaruinn, near L. Erisort by *Chater* and *Pankhurst*, 1980. 1, 6, 7, 10, 21

**S. rupicola** Lebel ex Le Jolis
Rock Sea-spurrey.
Corran na Creige.
Rocks near the sea.
Native.
North Uist, Lochmaddy *Shoolbred* 1894 (BM); there are also literature records from South Uist, (Harrison, 1941a) and Benbecula (Shoolbred, 1895).

South Uist, Benbecula and North Uist.   16, 18, 19.

**S. marginata** (DC) Kittel
*S. media* (L.) C. Presl
Greater Sea-spurrey.
Corran Mara Mòr.
Native.
Salt marsh and shingle.
Barra to Lewis.   1, 3, 4, 12, 15, 16, 19, 21.

**S. marina** (L.) Griseb.
*S. salina* J. & C. Presl
Lesser Sea-spurrey.
Corran Mara Beag.
Native.
Salt marsh, shingle and peat cuttings.
Barra to Lewis and North Rona and Sula Sgeir.   1–3, 8, 10, 12, 15–19, 21, 25, 29.

Herniaria ciliolata Melderis
This is a rare plant of Cornwall and the Channel Islands. It was recorded by *J. W. H. Harrison* from South Uist, golf course at Kildonan. There is material, named as Hernaria glabra L. var. ciliata by Pugsley, at K collected 1939. The species is very unlikely from this locality, and has not been seen there again.

Illecebrum verticillatum L.
A rare plant of the south and west, recorded from Barra, sandy lochside near Borve in 1939 by J. W. H. Harrison (K, BM). There is also a record from Eriskay (Goodrich-Freer, 1902). Requires confirmation; the habitat in Barra is now quite unsuitable.

*PORTULACACEAE*
**Montia** L.
**M. fontana** L.
Blinks.
Fliodh Uisge.**
Native
Ditches and flushes.
Common.
Mingulay to Lewis.
Little of the material has been determined to subspecies but the following have been determined by *Walters*:
**M. fontana** L. subsp. **fontana**
Lewis, Uig, Carnish *Bangerter and Crabbe* 1939 (E); Harris, Scalpay *J. W. Campbell* 1939.
1–10, 12–25, 28, 29.

**M. fontana** L. subsp. **chondrosperma** (Fenzl) Walters
Barra, Borve *Campbell* 1978 (BM).

**M. fontana** L. subsp. **variabilis** Walters
Lewis, Uig, Mealista *Crabbe* 1939 (BM); St. Kilda *Gibson* 1889 (E).
From Berneray in the south to Lewis and Hirta.

**M. sibirica** (L.) Howell
*Claytonia alsinoides* Sims.
Pink Purslane.
Seachranaiche.*
Introduced.
Lewis, Stornoway Castle grounds, Cunningham (1978); North Uist, garden weed at Sidinish, *J. W. Clark*, 1977.   5, 16
Only two records, but might spread.

*CHENOPODIACEAE*
**Chenopodium** L.
**C. album** L.
Fat-hen.
Càl Slapach.
Doubtfully native.
Arable fields, gardens and waste places.
Mingulay to Lewis.   1, 3, 11, 12, 15, 16, 18–24.

**C. opulifolium** Schrader ex Koch & Ziz
Grey Goosefoot.
Only recorded by *H. H. Harrison*, Outer Isles (Harrison, 1939) and Vatersay (J. W. H. Harrison, 1941a). Probably represents a form of the previous species and in the absence of specimens requires confirmation.

**Beta** L.
**B. vulgaris** L.
Subsp. **maritima** (L.) Thell.
*B. maritima* L.
Sea Beet.
Biatas Mara.
Native.
Seashores.
Rare. Pabbay, field record *Cheke*, 1954; North Uist, Baleshare, *Murray*, 1976 (BM).   16, 23

**Atriplex** L.
**A. littoralis** L.
Grass-leaved Orache.
Native.
J. W. H. Harrison (1941a) states 'Only noted on Bagh Scar, Vatersay and on the adjacent island of Flodday.' In the absence of specimens and the lack of any records from the west coast of Scotland this certainly requires confirmation.

**A. patula** L.
Common Orache.

Praiseach Mhìn.
Doubtfully native.
Benbecula to Lewis
A few specimens have been determined by *Wilmott* as **A. patula** L. var. **erecta** (Hudson) Lange, this seems to be just part of the normal variation of the species.   3, 6, 8, 11, 12, 15, 16, 18, 26.

**A. prostrata** Boucher ex DC.
(*A. hastata* auct., non L.)
Spear-leaved Orache.
Praiseach Mhìn.
Native.
Shoreline and gravelly places.
There are specimens from South Uist to Lewis and literature records that extend the distribution to Hirta and south to Barra.   1, 8, 12, 16, 19–21, 23, 28, 29.

**A. glabriuscula** Edmondston
Babington's Orache.
Praiseach Mhìn.
Native
Sandy shorelines.
Barra to Lewis, Hirta, North Rona.   1–3, 6–8, 10–14, 16, 18–23, 25, 28.

**A. praecox** Hülphers
Early Orache.
Native.
On usually rather muddy sand and shingle. First recognized in the British Isles by Taschereau (1985). It most closely resembles **A. patula** but differs in the bracteoles being little longer than the seed. There are two records confirmed by Taschereau but it is probably in other places.
North Uist, Carinish *Clark* 1978 (MANCH); South Uist, Hartavagh, *Pankhurst*, 1979 (BM). 16, 19.

**A. laciniata** L.
Frosted Orache.
Praiseach Mhìn Airgeadach.
Native.
Sandy shorelines.
Barra to Lewis.   3, 9, 12, 15, 16, 18, 19, 21, 29.

**Suaeda** Forskål ex Scop.
**S. maritima** (L.) Dumort.
Annual Sea-blite.
Praiseach na Mara.
Native.
Salt marshes.
From Berneray to Lewis.   1, 4, 15, 16, 18, 19, 21

**Salsola** L.
**S. kali** L.
Prickly Saltwort.
Lus an t-Salainn.**
Native.
Sandy Shorelines.
Barra, Eoligarry *Edin. Univ. Biol. Soc.* 1937; Benbecula, sandy shore near Croagary (presumably Creagorry) *Shoolbred* 1894. *J. W. H. Harrison* recorded it from Eriskay and Pabbay and Macgillivray (1830) from St. Kilda (Hirta).   16, 18, 20, 21, 23, 28.

**Salicornia** L.
**S. europaea** L.
Glasswort.
Lus na Glainne.**
Native.
Saltmarsh.
Rare. Barra, near Brevig *Edin. Univ. Biol. Soc.* 1936 and  *Campbell* 1978; Conacher (1980) also recorded it from Barra; Benbecula and North Uist (*Shoolbred*, 1895) and *J. W. H. Harrison* recorded it from Great Bernera, Benbecula (Creagorry) and from South Uist. 4, 16, 18, 19, 21

S. ramosissima J. Woods
In Harrison (1941a) this is recorded from South Uist, Benbecula and North Uist, there are also records for *S. stricta* Dumort., *S. gracillima* (Towns.) Moss and *S. prostrata* Pallas. However, anything other than **S. europea** requires confirmation.

**Note**
Material from Creagorry, Benbecula was separated as *S. scotica* Wilmott (unpublished) but the material was collected in November 1935 and is quite indeterminable.
A report on saltmarshes (Law and Gilbert, 1986) gives record for *Salicornia* agg. from zones   1, 2, 4, 12, 16–19.

*TILIACEAE*
**Tilia** L.
**T. × europaea** L.
(*T. × vulgaris* Hayne)
Lime.
Teile.
Has been occasionally planted, as in the grounds of Stornoway Castle.   5.

*MALVALES*
*MALVACEAE*
**Malva** L.
**M. moschata** L.

Musk Mallow.
Lus nam Meall Mòra.
Denizen.
Benbecula, outside cottage *J. W. Campbell*
1947 (BM).   18.

GERANIALES
LINACEAE
**Linum** L.
**L. catharticum** L.
Fairy Flax.
Lìon nam Ban-sìdh.
Native.
Common on machair, in grassland and on
moorland.
Mingulay to Lewis.   1, 3–7, 11–13, 15–24, 29.

**Radiola** Hill
**R. linoides** Roth
Allseed.
Lus Meanbh Meanganach.*
Native.
Rare on sandy ground.
This decreasing species is recorded from
Barra to Lewis but may be overlooked.   1–3,
6, 16–21.

GERANIACEAE
**Geranium** L.
**G. pratense** L.
Meadow Crane's-bill.
Crobh Preachain an Lòin.
Escape.
Rare. South Uist; Bornish, machair, Harrison
(1948); North Uist, near farmhouse at
Balelone, a var. **flore-pleno** is said to have
come from Coll, *J. A. Clark* (BM); South Harris,
Kyles Lodge, Harrison (1948).   12, 16, 19

**G. sylvaticum** L.
Wood Crane's-bill.
Crobh Preachain Coille.
Introduction.
Lewis, grounds of Stornoway Castle,
Harrison (1948) and Cunningham (1978).   5

**G. dissectum** L.
Cut-leaved Crane's-bill.
Crobh Preachain Geàrrte.
Native.
Machair and fields.
Barra, South and North Uist.   9, 12, 16, 19,
21.

**G. rotundifolium** L.
Round-leaved Crane's-bill.
Given in Harrison et al. (1941a) as 'Around

East Loch Tarbert, North and South Harris';
requires confirmation.

**G. molle** L.
Dove's-foot Crane's-bill.
Crobh Preachain Mìn.
Native.
Sandy cultivated ground, waste places.
Mingulay to Lewis.   1, 3, 12–16, 18–24, 29.

**G. pusillum** L.
Small-flowered Crane's-bill.
Only recorded by Harrison (1941a, 1956)
from South Uist, near Stoneybridge 1941;
and Lewis, Stornoway 1956. The records
may be errors for the last.   ?1, ?19.

**G. robertianum** L.
Herb-Robert.
Lus an Ròis, Ruideal.
Native.
Rocky places.
Very   rare,   specimens   from   Harris,
Amhuinnsuidhe *Duncan* 1896 (BM); South
Uist, Lochskiport Pier *Chorley* 1983 (BM).
Literature records from J. W. H. Harrison
(1941a) from South Uist, Allt Volagir ravine
and east coast near Loch Skiport.   9, 19.

**Erodium** L'Hérit.
**E. cicutarium** (L.) L'Hérit.
Common Stork's-bill.
Gob Corra.
Native
Subsp. **cicutarium** and subsp. **dunense**
Andreas are both recorded.
Dunes, machair arable fields and roadsides.
Barra to Lewis.   1, 12, 14–16, 18–24, 29.

OXALIDACEAE
**Oxalis** L.
**O. acetosella** L.
Wood-sorrel.
Feada-coille, Biadh nan Eòinean.
Native.
Cliffs, gorges, loch edges and moors.
Barra to Lewis.   1–3, 5, 7, 9–12, 16–19, 21.

SAPINDALES
ACERACEAE
**Acer** L.
**A. pseudoplatanus** L
Sycamore.
Craobh Pleantrainn.
Planted.
In woods and plantations as at Barra,
Northbay *Edin. Univ. Biol. Soc.* (E), there is a

photograph in Watson (1939b); South Harris, Rodel (*Shoolbred*, 1895). 5, 7, 12, 16, 19, 21.

**A. campestre** L.
Field Maple.
Craobh Mhalpais.
Planted.
South Harris, Tarbert area, field record *Coulson* 1971; Lewis, grounds of Stornoway Castle, Cunningham (1978). 5, 12

*HIPPOCASTANACEAE*
**Aesculus** L.
**A. hippocastanum** L.
Horse-chestnut.
Craobh Geanm-chnò Fhiadhaich.
Planted.
Barra, field record Northbay, *Chater* 1983; Lewis, Stornoway Castle grounds, Cunningham (1978). 5, 21

CELASTRALES
*AQUIFOLIACEAE*
**Ilex** L.
**I. aquifolium** L.
Holly.
Cuileann.
Native.
Cliffs and gullies, also possibly planted.
Harris and Lewis. 3, 5, 7, 9, 12

*CELASTRACEAE*
**Euonymus** L.
**E. europaeus** L.
Spindle.
Feoras.
Planted.
Lewis, Stornoway Castle grounds, Cunningham (1978). 5

ROSALES
*LEGUMINOSAE*
**Cytisus** L.
**C. scoparius** (L.) Link
*Sarothamnus scoparius* (L.) Wimm. ex Koch
Broom.
Bealaidh.
Probably always planted as at Lewis, Stornoway, Castle Park *Campbell* (BM), at Eishken and at Harris, Tarbert, The Manse *Shoolbred* (1895) and at Sidinish, North Uist *J. W. Clark*. 1, 5, 7, 9, 16

**Ulex** L.
**U. europaeus** L.
Gorse.
Conasg.

Introduced or naturalised.
Roadsides, plantations.
Barra, the Uists, Harris and Lewis   1–3, 5–7, 9, 12, 16, 19, 21, 27

**Vicia** L.
**V. hirsuta** (L.) Gray
Hairy Tare.
Peasair an Arbhair.
Status uncertain.
Rare, only a single record in literature; St Kilda, *Petch* near Factors house, Petch (1933a). 28

**V. cracca** L.
Tufted Vetch.
Peasair nan Luch.
Native
Grassland and roadsides.
Berneray and Mingulay to Lewis   1–4, 6, 7, 9, 10, 12, 14–16, 18–22, 24

**V. orobus** DC.
Wood Bitter-vetch.
Peasair Shearbh.
No specimens seen but recorded by Harrison (1941a) from Eriskay 'cliffs on east coast'. The status of this record is uncertain.

**V. sylvatica** L.
var. **sylvatica**
Wood Vetch.
Peasair Coille.
Native.
Obbe, Harris *Duncan* (BM); Glen of Rodell, South Harris (Macgillivray, 1830) 12

var. **subrotundata** A. Bennett
North coast of Scarp *Duncan* (type BM).   8

**V. sepium** L.
Bush Vetch.
Peasair nam Preas.
Native.
Rocky outcrops and fields.
Mingulay to Lewis, St. Kilda, Soay.   4, 12–14, 16, 21, 23, 24, 28

**V. sativa** L.
subsp. **sativa**
Common Vetch.
Peasair Chapaill.
Denizen
There is a single specimen from Crowlista, Lewis *M. S. Campbell* (BM) which matches the modern concept of this taxon.   3

subsp. **nigra** (L.) Ehrh.
*V. angustifolia* L.
Narrow-leaved Vetch.
Peasair nan Coilleag.
Denizen.
Newton, North Uist *M. S. Campbell* (BM);
North Uist *Balfour* (E)    16
All records:  3,16,19

**V. lathyroides** L.
Spring Vetch.
Native.
Calcareous turf.
One record only. Lewis, Loch na Berie,
*McKean* (E), 1985.    3

**Lathyrus** L.
**L. pratensis** L.
Meadow Vetchling.
Peasair Bhuidhe.
Native.
Roadsides.
Berneray and Mingulay to Mid-west Lewis.
2–6, 11, 12, 14, 16, 18, 19, 21–24

**L. montanus** Bernh.
Bitter-vetch.
Cairt Leamhna.
Native.
Specially on more acid soils.
South Uist and Harris and Lewis    2–5, 7, 9,
12, 19

**Ononis** L.
**O. repens** L.
Restharrow.
Sreang Bogha.
Native.
Rare. Vatersay (Harrison, 1941a) and
Conacher (1980); Ensay, on dunes *J. W. H.
Harrison* 1948 (K) and Harrison (1956)  14,
22

**Medicago** L.
**M. sativa** L.
Lucerne.
Escape.
Only specimen is from North Uist, Newton
*J.W.Campbell* (BM) 1937; literature record
from South Harris, Borve, Harrison (1954a).
12, 16

**M. lupulina** L.
Black Medick.
Dubh-mheidig.**
Native or introduction.
Harris and the Uists, only specimen is

Newton, North Uist *M. S. Campbell* 1936
(BM), confirmed by field record 1979; S.
Harris, Rodel 1979, field record; there are
literature records from Lochmaddy, North
Uist (Harrison et al., 1942a) and Lochboisdale,
South Uist, (Harrison et al. 1942b), and
Stoneybridge, South Uist (Harrison et al,
1939).   12, 16, 19

**Trifolium** L.
**T. repens** L.
White Clover.
Seamrag Bhàn.
Native and denizen.
Machair and grassland.
Probably in all zones.   1–25, 27, 28

**T. hybridum** L.
Alsike Clover.
Introduction.
Only two specimens seen; South Uist, 1888
*Somerville*; Enaclete, Lewis *M. S. Campbell*
(BM). Literature records from Lochmaddy
(Harrison et al., 1942a) and Stornoway, Lews
Park, Lewis (Copping, 1977).   3, 5, 16, 19

**T. fragiferum** L.
Strawberry Clover.
Native.
Machair.
A record from South Uist by *Meinertzhagen*
(1947) was confirmed by Pankhurst in 1983;
machair north of Loch Bee, South Uist
*Pankhurst* (BM)   19

**T. campestre** Schreber
Hop Trefoil.
Seamrag Bhuidhe.
Doubtfully native.
Tracksides.
Barra to South Uist, only a single specimen
seen on a roadside, Barra 1951 *Goodwin and
Goodwin* (GL); there are literature records
from Barra, (Shoolbred, 1895) and from South
Uist (Harrison et al., 1939) and Harrison
(1941a).   19, 21

**T. dubium** Sibth.
Lesser Trefoil.
Seangan.
Native.
Roadsides.
Barra to Lewis   1–3, 9, 12, 14, 16, 18, 19, 21,
23

**T. pratense** L.
Red Clover.

Seamrag Dhearg.
Native and persisting after sowing.
Machair and grasslands.
Common, from Berneray and Mingulay to
Lewis.   1–7, 9–24, 27–29

**T. medium** L.
Zigzag Clover.
Seamrag Chrò-dhearg.
Possibly native.
Banks and roadsides.
Barra to Lewis   1–3, 9?, 12, 14, 19, 21

**Anthyllis** L.
**A. vulneraria** L. ·
Kidney Vetch.
Cas an Uain.
Native.
Machair and sandy banks.
According to *Akeroyd*, most of the material
corresponds to subsp. **lapponica** (Hyl.) Jalas,
but some of the collections tend towards
subsp. **vulneraria**; **A. vulneraria** var.
**maritima** is also recorded.
Berneray and Mingulay to Lewis.   1–4, 6, 9,
12, 13, 15, 16, 18–24, 29

**Lotus** L.
**L. corniculatus** L.
Common Bird's-foot-trefoil.
Barra-mhìslean, Peasair a' Mhadaidh-Ruaidh.
Native.
Machair, roadsides and dry banks.
The var. **villosus** is recorded from North
Uist.
Probably in all zones.   1–22, 24, 27, 29

**L. uliginosus** Schkuhr
Greater Bird's-foot-trefoil.
Barra-mhìslean Lèana.
Native.
Dry ditches.
There are specimens from South and North
Uist and Lewis (Stornoway Castle); South
Harris, Rodel, field record.   1, 5, 12, 16, 19

*ROSACEAE*
**Spiraea** L.
**S. alba** Du Roi var. **latifolia** (Aiton) Dippel
(*S.latifolia* (Aiton) Borkh., *S.salicifolia* L. var.
*latifolia* Aiton)

Bridewort.
Introduction
Lewis, Miavaig, *Crabbe* (BM); South Harris,
near Borve, Harrison (1954a).   3, 12
The only specimen has the almost glabrous
leaves and inflorescence and ovate leaves of
S. alba, but was originally determined as S.
salicifolia L.

**Filipendula** Miller
**F. ulmaria** (L.) Maxim.
Meadowsweet.
Cneas Chù Chulainn.
Native.
Marshes and ditches.
Berneray and Mingulay to Lewis and the
Shiant Is.   3–7, 9–16, 18–24, 27, 28

**Rubus** L.
**R. saxatilis** L.
Stone Bramble.
Caor Bad Miann.
Native.
Uncommon. In ravines and by streams on
rocks.   1, 3, 9, 12, 18, 19, 22

**R. idaeus** L.
Raspberry.
Subh-craoibh.
Native.
On rocks and cliffs.
Scarce.   3, 5, 7, 12, 19, 21, 23, 27

**R. spectabilis** Pursh
Canadian Raspberry.
Subh-craoibh Ruiteach.
Introduced and naturalised.
Lewis, in the grounds of Stornoway Castle,
Cunningham (1978).   5

**R. fruticosus** L. agg.
Bramble.
Dris (for the bush), Smeur (berry).
Native.
In ravines of streams, on sea cliffs, islands
and margins of lochs. Apparently restricted
by grazing and exposure, rare in the north
but becoming frequent in the south, as in
South Uist and Barra.   5, 7, 9, 10, 12, 15, 16,
18–23

The following account of the microspecies of R. fruticosus is based on the recent monograph
by Edees and Newton (1988). Only records which are supported by herbarium specimens
which have been determined by one or both of the above experts are included. For accurate
determination, the specimen should consist of a stem piece (in its first year of growth) and a
panicle. This has meant that a proportion of the plants which have been found cannot be

named because the material is too poor. August is the best month for collecting. The taxonomy of the group has so changed with time that the early records using older names can often not be equated with modern species unless there is a specimen. It is interesting to note that the Outer Hebrides has roughly twice as many native species as does Skye, and that the Orkneys and Shetland have no native species at all. The following key uses vegetative characters as far as possible. A short petiolule is one that is up to and including one quarter the length of the leaflet. By 'unequally serrate' is meant that some teeth are much wider and longer than others.

1   Stem with many long hairs...........................................................................2
    Stem with few or no long hairs......................................................................7

2   Stem prickles unequal, with sparse to dense stalked glands .....................................3
    Stem prickles equal, stem without stalked glands ...............................................5

3   Terminal leaflet equally serrate, anthers hairy .................................... *mucronulatus*
    Terminal leaflet unequally serrate, anthers glabrous ...........................................4

4   Leaves glabrous above, terminal leaflet petiolule fairly long, some panicle prickles straight and some curved, sepals long-pointed ...................................... *dasyphyllus*
    Leaves sparsely hairy above, terminal leaflet petiolule short, panicle prickles all more or less straight, sepals short-pointed ...................................... *adenanthoides*

5   Stem not furrowed, sparsely felted, anthers hairy.................................... *leptothyrsus*
    Stem distinctly furrowed, not felted, anthers glabrous...........................................6

6   Leaflets flat, not undulate or plicate, terminal leaflet equally serrate, sepals short-pointed, carpels glabrous...................................... *septentrionalis*
    Leaflets concave, undulate or plicate, terminal unequally serrate, sepals long-pointed, carpels hairy...................................... *incurvatus*

7   Lateral leaflets subsessile .........................................................................8
    Lateral leaflets distinctly stalked .................................................................9

8   Leaflets flat, not undulate or plicate, terminal leaflet cordate at base, petals flat, stamens about equalling styles, carpels glabrous...................................... *latifolius*
    Leaflets concave, undulate or plicate, terminal leaflet entire or emarginate at base, petals concave, stamens longer than styles, carpels hairy...................................... *hebridensis*

9   Leaves grey or white below ......................................................................10
    Leaves green below ..............................................................................17

10  Stem sparsely to densely glandular ...............................................................11
    Stem without stalked glands .....................................................................13

11  Tip of terminal leaflet cuspidate, margins equally serrate, carpels hairy.............. *polyanthemus*
    Tip of terminal leaflet acuminate, margins unequally serrate, carpels glabrous....................12

12  Stem prickles often short and unequal, terminal leaflet rhombic, petiolule short, panicle rachis markedly flexuose (zig-zag)...................................... *adenanthoides*
    Stem prickles mostly longer and subequal, terminal leaflet usually ovate, petiolule fairly long, panicle rachis slightly flexuose ...................................... *radula*

13  Leaves glabrous above, panicle prickles all curved ..............................................14
    Leaves sparsely to distinctly hairy above, at least some panicle prickles more or less straight ........15

14  Stem not felted or pruinose, not furrowed, sepals long-pointed or leafy-tipped, styles not coloured ...................................... *nemoralis*
    Stem sparsely or densely felted, pruinose, distinctly furrowed, sepals short-pointed, styles coloured ...................................... *ulmifolius*

15  Terminal leaflet equally serrate, cuspidate at tip (especially in the panicle), panicle stalked glands few ...................................... *polyanthemus*
    Terminal leaflet unequally serrate, acuminate at tip, panicle stalked glands absent .................16

16  Stem not furrowed, terminal leaflet obovate, panicle prickles all more or less straight, sepals short-pointed...................................... *dumnoniensis*
    Stem distinctly furrowed, terminal leaflet ovate or roundish, some panicle prickles straight and some curved, sepals long-pointed ...................................... *incurvatus*

17  Stem erect, sepals green ...................................... *fissus*
    Stem procumbent or arching, sepals greyish ......................................................18

## Section Rubus

**R. fissus** Lindley
Native.
South Uist, Crossdougal,*Chater and Pankhurst*,
1980 (BM), and L. Eynort (Harrison, 1941a).
19

### Series Sylvatici (Mueller) Focke

**R. errabundus** W. C. R. Watson
Native.
South Uist, Howmore, *J. W. Clark*, 1975 (E)
and Grogarry, *Wilmott*, 1947 (BM).   19

**R. leptothyrsus** G. Braun
(*R. danicus* Focke)
Native.
South Uist, North Glendale, *Chater &*
*Pankhurst*, 1980 (BM); Benbecula, *Shoolbred*,
1894 (E); South Harris, Grosebay, *Chater &*
*Pankhurst*, 1980 (BM).   12, 18, 19

### Series Rhamnifolii (Bab.) Focke

**R. ebudensis** A. Newton
Native.
Widespread from South Uist to south Lewis.
Also known from Skye. South Uist, L.
Boisdale, *Crompton*, 1964 (CGE), North
Glendale, *Chater & Pankhurst*, 1980 (BM);
Benbecula, *J. W. H. Harrison* 1940 (CGE);
South Harris, Leverburgh and Ardvey,
*Chater & Pankhurst*, 1980 (BM); Lewis,
Eishken *Chater & Pankhurst*, 1980 (BM).
Previous records (in brackets) of **R. gratus**
probably refer to this before it was recog-
nised as a new species (Newton, 1988).   7,
(9), 12, (16), 18, 19

**R. dumnoniensis** Bab.
Native.
Barra, Castlebay, *J. W. Clark*, 1978 (E);
Benbecula, *J. W. H. Harrison*, 1940 (CGE).
Unconfirmed record from South Uist.   18,
(19), 21

**R. incurvatus** Bab.
Native.
North Harris, Kyles Scalpay, *Chater &*
*Pankhurst*, 1980 (BM); unconfirmed records
from North Uist and Benbecula.   9, (16, 18)

**R. nemoralis** P. J. Mueller
(*R.selmeri* Lindeb.)
Native.
Barra, North Bay, *Pankhurst*, 1983 (BM);
South Harris, Borve, *Chater & Pankhurst*,
1980 (BM); Lewis, grounds of Stornoway
Castle, *Chater & Pankhurst*, 1980 (BM). Un-
confirmed records from Benbecula.   5, 12,
(18), 21.

**R. polyanthemus** Lindeb.
Native.
Frequent in Barra, the Uists and Benbecula;
also on Berneray.   15, 16, 18, 19, 21

**R. septentrionalis** W. C. R. Watson
Native.
Frequent, South Uist to Lewis.   5, 7, 9, 12,
16, 18, 19

Series Discolores (Mueller) Focke

**R. ulmifolius** Schott
Introduced.
South Harris, Borve Lodge, in walled garden, field record *Pankhurst* 1980.   12

Series Mucronati (Focke) H. E. Weber

**R. mucronulatus** Boreau
Native.
Quite frequent, South Uist to Lewis.   7, 12, 16, 18, 19, 20

Series Radulae (Focke) Focke

**R. adenanthoides** A. Newton
Native.
Rare, and not seen recently. South Uist, near Eynort, *J. W. H. Harrison*, 1939 (CGE); Benbecula, ibid. 1940 (CGE).   18, 19

**R. radula** Weihe ex Boenn.
Native.
Vatersay, *Cannon & Cannon*, 1983 (BM); Barra, Castlebay, *Shoolbred*, 1894 (NMW); Lewis, Eishken, *Pankhurst & Chater*, 1980 (BM) and Stornoway Castle, *J. W. Clark*, 1975 (E). Unconfirmed record from Benbecula. 5, 7, (16), 18, 21, 22

An unnamed robust and glandular bramble with large white flowers belonging to this section was collected in 1980 at Gravir, Park on the north side of L. Odhairn (*Pankhurst & Chater* 344, BM). It is quite unlike any of the recorded species and is most unlikely to be a hybrid, since bramble hybrids are usually depauperate plants. It does not correspond to any named species, but might have spread from Ireland.

Series Hystrices Focke

**R. dasyphyllus** (Rogers) E. S. Marshall
Doubtfully native.
Barra, near Brevig, *Pankhurst* 1983 (BM); Lewis, in grounds of Stornoway Castle, *Pankhurst & Chater*, 1980 (BM).   5, 21

Section Corylifolii Lindley

**R. hebridensis** E. S. Edees
Native.
South Uist, near Crossdougal, *Wilmott*, 1947 (BM); South Harris, *Shoolbred* 1894 (MANCH).

A record of R. conjungens by Balfour and Babington (1842) at Rodel, South Harris, in 1841 may refer to this species. Not seen since 1947. Scarce in the vice-county, in spite of its name. See Edees (1975).   12, 19

**R. latifolius** Bab.
Native.
Island of Sandray, Barra, *J. W. H. Harrison*, 1939 (CGE).   23

**Potentilla** L.
**P. palustris** (L.) Scop.
Marsh Cinquefoil.
Còig-bhileach Uisge.
Native.
Marshes and wet ditches.
Barra to Lewis   1–4, 6, 7, 12, 13, 15–19, 21–23

**P. anserina** L.
Silverweed.
Brisgean.
Native.
Machair, sand-dunes, grasslands, sea-shore and waste places.
Subsp. **egedii** (Wormsk.) Hiitonen is recorded from Shetland but not as yet from the Outer Isles. Its leaves have 7–15 leaflets and are glabrous to sparsely pubescent below, as opposed to 15–25 leaflets which are silvery-sericeous below in subsp. **anserina**. Most islands and outliers from Berneray to Lewis and from St. Kilda, North Rona and the Monach Is. to the Shiant Is.   1–25, 27–29

**P. erecta** (L.) Räuschel
Tormentil.
Cairt Làir.
Native.
Moorland, machair (grassland), and dry banks.
Berneray and Mingulay to Lewis and St. Kilda to Shiant Is.   1–10, 12, 13, 15–24, 27–29

P. reptans L.
Creeping Cinquefoil.
Còig-bhileach.
There are field records for North Uist and Great Bernera, but no voucher specimens. Confirmation is required.

**Fragaria** L.
**F. vesca** L.
Wild Strawberry.
Subh-làir.

Native.
Barra, South Uist, South Harris and Lewis
(Stornoway Castle).   1, 5, 12, 18, 21

**Geum** L.
**G. urbanum** L.
Wood Avens.
Machall Coille.
Possibly introduced.
Rare. Lewis, woods of Stornoway Castle,
Harrison (1948). Not reported recently, and
needs confirming.   5

**G. rivale** L.
Water Avens.
Machall Uisge.
Native.
Only one specimen seen; Creagan Leathan,
Ulladale, North Harris *Wilmott* (BM). The
Atlas shows another record from near
Tarbert. Curiously rare.   9

**Agrimonia** L.
A. eupatoria L.
Agrimony.
Geur-bhileach.
There are two records for this plant from
South Uist, but both specimens have been
redetermined as the next species.

**A. procera** Wallr.
*A. odorata* auct. non Miller
Fragrant Agrimony.
Geur-bhileach.
Native.
North side of Lochboisdale, South Uist. First
recorded (as **A. eupatoria**) *Somerville* 1888
(BM) also seen by *J. W. Campbell* 1947 (BM)
19

**Alchemilla** L.
**A. alpina** L.
Alpine Lady's-mantle.
Trusgan, Meangan Moire.
Native.
Restricted to the mountain massif of North
Harris from Ceartoval in the West to
Clisham in the East. It is surprising that it is
not known from the high ground in the
Uists.   9

**A. glabra** Neygenf.
Smooth Lady's-mantle.
Fallaing Moire.
Native.
Vatersay to Harris.   12, 19, 21, 22

A. vulgaris L. *sensu lato*. Records almost
certainly represent A. glabra.

**A. wichurae** (Buser) Stefánsson
There are literature records for this from
Hecla, South Uist and Moor Hill, Berneray,
Harris *J. W. H. Harrison* but these are almost
certainly the last species.

**Aphanes** L.
**A. arvensis** L.
Parsley-piert.
Spìonan Moire.
Native.
Surprisingly    restricted;    South    Harris,
Benbecula, North Uist, Scarp and Lewis
near Stornoway.   1, 5, 8, 12, 16

**A. inexpectata** Lippert
(*A. microcarpa* auctt. non (Boiss. & Reuter)
Rothm.)
Native.
Scarce, no voucher specimens. North Uist,
Newton, *Pankhurst*, 1979; South Harris, near
Tarbert, ibid.; Eishken, *Pankhurst & Chater*,
1980; Lewis, Stornoway Castle, ibid. and
Harrison (1957), Carloway area, *J. F. M.
Cannon*, 1979.   2, 5, 7, 12, 16

**Rosa** L.
This account was prepared with the assistance of the Rev. G. Graham. Roses are
a difficult genus taxonomically because of
frequent hybridisation with unequal sharing
of chromosomes. Current thinking is that
many of the names which were used for
hybrid forms in the past are now untenable,
since only a small part of the full range of
actual variation was accounted for, and
because the names were used inconsistently
in different parts of the country. Hence in
this account, we only accept records for the
principal species. However, Rose specialists
were not so cautious in the earlier part
of this century, and created many names
which are no longer in use. Harrison and
Bolton (1938) published what, at the time,
was a definitive account of Roses in the
Hebrides, but we cannot now approach the
subject with such confidence. Early records
are found in Shoolbred (1895) and Prof.
Harrison and his colleagues published a
great many records (Clark and Harrison,
1940; Harrison, 1941a and 1956; Harrison et
al., 1942a), but generally without voucher
specimens. When the Rose specimens in the
herbarium of *W. A. Clark* were examined

they were found to be so badly decayed that none of them could be named.

Other than *R. pimpinellifolia* the Roses in the islands are typically rather uncommon and found as small bushes on cliffs and on the sides of river ravines, in contrast with their characteristic southern English woodland habitat, which is more or less lacking in the Islands. Most of the plants seen during our surveys were depauperate, which may explain why relatively few collections seem to have been made, and why quite a proportion of these have not been named. The distribution given below relates to all species (other than *R. pimpinellifolia*). 2–5, 7–14, 16–24.

**R. pimpinellifolia** L.
Burnet Rose.
Ròs Beag Bàn na h-Alba.*
Native.
Recorded from Mingulay and several places on Barra (Ben Erival, Ben Scurrival and Tangusdale) with one record from South Harris (Borvebeg Burn). 12, 21, 24

**R. rugosa** Thunb. .
Planted.
Barra, field record from Halaman Bay area, 1978, comm. *Murray*. 21

**R. canina** L.
Dog-rose.
Ròs nan Con.
Native.
Cliffs, river ravines, streamsides.
Widely distributed but scarce. 4, 5, 7, 10–12, 14, 16–24

**R. afzeliana** Fries
(*R. dumalis* auct.)
Glaucous Dog-rose.
Native.
Distribution similar to the above. 3–5, 8, 9, 11, 12, 16, 20–22, 24

R. tomentosa Sm.
A specimen at the BM reputed to be this species from South Harris (Rodel, *W. S. Duncan*) has been redetermined by *Graham* as R. sherardii; and specimens from the Lee Hills in North Uist, and from Taransay, coll. *Shoolbred* as R. mollis.

**R. sherardii** Davies
Sherard's Downy-rose.
Native.

Widely distributed, but mostly in the southern islands. 4, 11, 12, 14, 16, 18–21 Harrison (1938a) published var. **cookei** from the south of South Uist. It resembles **R. pimpinellifolia** in its armature, and could be mistaken for that species. 4, 16, 19

**R. sherardii** × **R. rubiginosa**
(*R. sherardii* var. *suberecta* (Ley.) H.-Harr.) *Graham* (pers.comm.) confirms a specimen collected by *J. W. Clark* at Hoe Beg in North Uist, 1983 (SUN) and states that var. **suberecta** is in fact a hybrid of the above parentage. Harrison (1941a) also records it from Sandray, Barra, South Uist, Benbecula and Great Bernera. 4, 16, 18, 19, 21, 23

**R. mollis** Sm.
(*R. villosa* L.)
Soft Downy-rose.
Native.
Mainly in the south from South Harris to Barra, apart from one record in Lewis (between Carloway and Breasclete). 2, 7, 11, 12, 16, 19, 21, 22

R. rubiginosa L. is recorded in the BSBI Atlas. *Preston* (pers.comm.) suggests that these are equivalent to records by Harrison (1941a) for R. sherardii var. suberecta (see above).

**Prunus** L.
**P. avium** (L.) L.
Wild Cherry.
Geanais.
Planted.
Barra, field record BSBI meetimg 1978; Lewis, Stornoway Castle, Cunningham (1978). 5, 21

**P. spinosa** L.
Blackthorn.
Sgitheach Dubh, Preas nan Airneag.
Planted.
South Harris, Borve, Harrison (1954b); South Uist, planted in hedgerow at Grogarry House, field record 1983. 12, 19

**Cotoneaster** Medicus
**C. simonsii** Baker
Himalayan Cotoneaster.
Planted.
South Harris, near Borvemor, Harrison (1954a). 12

**C. horizontalis** Decne.
Wall Cotoneaster.

Cotaineaster Balla.**
Introduction
First record, cliffs near Hotel, Rodel, South
Harris *Chater* 1980.    12

**C. microphyllus** Wall. ex Lindley
Small-leaved Cotoneaster.
Cotaineastar Mion-dhuilleagach.**
Introduction.
First record from Lews Park, Stornoway,
Lewis *Copping* (1977), also reported by
Cunningham (1878).    5

**Crataegus** L.
**C. monogyna** Jacq.
Hawthorn.
Sgitheach.
Native and planted.
Lewis, North and South Harris, Berneray
and South Uist. Planted at Stornoway Castle
(Cunningham, 1978), Eishken, Rodel and
Grogarry.    5, 7, 9, 12, 15, 19

**Sorbus** L.
**S. aucuparia** L.
Rowan.
Caorann.
Native.
Cliffs and ravines.
Barra to Lewis    3–5, 7, 9–12, 16–21

**S. intermedia** (Ehrh.) Pers.
Swedish Whitebeam.
Planted.
Castle Park, Stornoway, Lewis *M. S.
Campbell* 1938 (BM) and again Cunningham
(1978).    5

**S. aria** (L.) Crantz var. **incisa** Reichenb.
Common Whitebeam.
Gall-uinnseann.
Planted.
Castle Park, Stornoway, Lewis *M. S.
Campbell* 1938 (BM) and again Cunningham
(1978).

*CRASSULACEAE*
**Sedum** L.
**S. rosea** (L.) Scop.
(*Rhodiola rosea* L.)
Roseroot.
Lus nan Laoch.
Native.
Mountain rocks and cliff ledges.
Berneray and Mingulay to Lewis and St
Kilda to Shiant Is.    1–6, 9, 12, 13, 16, 19–24,
27–29

**S. anglicum** Hudson
English Stonecrop.
Biadh an t-Sionnaidh.
Native.
Dry rocky places.
Berneray and Mingulay to Lewis and St
Kilda.    1, 3–7, 9–24, 27, 29

**S. acre** L.
Biting Stonecrop.
Grabhan nan Clach.
Native.
Sandy and rocky seashores.
Barra and South Uist to Lewis.    3, 9, 12, 13,
15, 16, 18–24, 29

**Crassula** L.
**C. tillaea** Lester-Garland
Mossy Stonecrop.
Isle of Barra *J. W. H. Harrison* 1939 (BM) the
specimen is as determined and the label
mentions Illecebrum verticillatum L. Both
these species are rare plants of South-west
Britain; the Crassula may be an introduction, but both records need independent
confirmation.

*SAXIFRAGACEAE*
**Saxifraga** L.
**S. stellaris** L.
Starry Saxifrage.
Clach-bhriseach Reultach.
Native.
Damp mountain rocks and cliff ledges.
Restricted to the tracts of higher ground,
South and North Uist, South and North
Harris and Lewis.    3, 5, 9, 12, 16, 19

**S. umbrosa** L.
Pyrenean Saxifrage.
Càl Phàdraig.
Introduction.
Goulaby Burn, North Uist *M. S. Campbell*
1936 (BM); Castle Park, Stornoway, Lewis
*Wilmott* 1946 (BM)    5, 16

**S. tridactylites** L.
Rue-leaved Saxifrage
Clach-bhriseach na Machrach.
Native.
Machair.
South Uist and Baleshare, North Uist.    16, 19

**S. rosacea** Moench
Irish Saxifrage.
Damp mountain ledges.
Status uncertain; a specimen exists labelled

from South Uist *J. W. H. Harrison* 1943 (BM) which appears to be this plant.

**S. hypnoides** L.
Mossy Saxifrage.
Clach-bhriseach Còinnich.
Native.
Restricted to the summit of Beinn Mhor, South Uist (Harrison et al., 1942b); *Polunin* 1944 (BM)

S. aizoides L.
Yellow Saxifrage.
Reported from a gully between Heaval and Hartaval on Barra by Harrison (1957). Not found by British Museum excursion in 1983, and no specimen has been seen.

**S. oppositifolia** L.
Purple Saxifrage.
Clach-bhriseach Purpaidh.
Native.
Cliffs, calcicole.
South Uist, North Harris, Lewis and St Kilda.   3, 9, 19, 28

**Chrysosplenium** L.
**C. oppositifolium** L.
Opposite-leaved Golden-saxifrage.
Lus nan Laogh.
Native.
South Uist, Harris and Stornoway, Lewis. 5, 12, 19

**Parnassia palustris** L.
Grass-of-Parnassus.
Reported by Gordon (1923) from South Uist, and from Kirkibost in North Uist (Pitkin et al., 1983) but without a voucher.   16, 19

*ESCALLONIACEAE*
**Escallonia** L.f.
**E. rubra** (Ruiz & Pavón) Pers. var **macrantha** (Hooker & Arnott) Reiche (*E. macrantha* Hooker & Arnott)
Escallonia.
Escape.
North Harris, on cliffs of East Loch Tarbert, Harrison (1954a).   9

*GROSSULARIACEAE*
**Ribes** L.
**R. nigrum** L.
Black Currant.
Dearc Dhubh, Raosar Dubh.
Probably denizen.
Skiport Road, near Grogarry, South Uist *Wilmott et al.* 1947 (BM)

SARRACENIALES
*DROSERACEAE*
**Drosera** L.
**D. rotundifolia** L.
Round-leaved Sundew.
Lus na Feàrnaich.
Native.
Bogs and wet heaths.
Berneray and Mingulay to Lewis and St. Kilda.   1, 3–7, 9–13, 15–25, 27, 28

**D. anglica** Hudson
Great Sundew.
Lus a' Ghadmainn.
Native.
Wetter parts of bogs.
Barra to Lewis.   1, 3–7, 9, 11, 12, 16, 18, 19, 21

**D.** × **obovata** Mert. & Koch
*D. anglica* × *rotundifolia*
In a few places from South Uist to Lewis, usually with the parents; also recorded from St Kilda. Probably under-recorded.   3, 5, 9, 12, 16, 19, ?28

**D. intermedia** Hayne
Oblong-leaved Sundew.
Dealt Ruaidhe.
Native.
Eriskay, Barra, Uists, South Harris and Lewis   1, 12, 16, 19, 21

MYRTALES
*LYTHRACEAE*
**Lythrum** L.
**L. salicaria** L.
Purple-loosestrife.
Lus na Sìochaint.
Doubtfully native.
Barra, South and North Uist and Uig, Lewis. 3, 16, 19, 21

**L. portula** (L.) D. A. Webb
*Peplis portula* L.
Water-purslane.
Flùr Bogaich Ealaidheach.*
Native.
Wet places.
Benbecula and the Uists and Lewis.   1, 2, 13, 16, 18, 19

*ELEAGNACEAE*
**Hippophae** L.
**H. rhamnoides** L.
Sea-buckthorn.
Planted.

North Uist, in gardens and by streams, *Shoolbred* (1899); South Harris, near Borve, Harrison (1954a).   12, 16.

*ONAGRACEAE*
**Epilobium** L.
**E. hirsutum** L.
Great Willowherb.
Seileachan Mòr.
Native.
Rare. Barra, field record Vaslain marsh, 1958; South Uist, Lochboisdale, Cooke (1944) reported in Vasculum 29:6 and Loch Ollay, Royal Botanic Garden, Edinburgh (1983); Benbecula, Loch Torcusay (ibid.); Great Bernera, Loch Sandavat, Harrison (1957). 4, 18, 19, 21

**E. parviflorum** Schreber
Hoary Willowherb.
Seileachan Liath.
Native.
Wet ditches and marshes.
Berneray and Mingulay to Harris and Lewis 5, 11–16, 18–21, 22, 24, 29

**E. montanum** L.
Broad-leaved Willowherb.
Seileachan Coitcheann.
Native.
Cliff ledges, or as a weed.
South Uist to Lewis.   3, 5–7, 12, 16, 18, 19

**E. ciliatum** Rafin.
*E. adenocaulon* Hausskn.
American Willowherb.
Seileachan Aimeireaganach.
Introduction.
South Uist, Harrison (1941a); Lewis, two records by the BSBI field meeting to Stornoway in August 1975; Lewis (Park) at Eishken.   5, 7, 16

**E. tetragonum** L.
*E. adnatum* Griseb.
Square-stemmed Willowherb.
Probable introduction.
Rare. Literature records from Fuday and Eriskay (Harrison et al, 1939). Only modern record from N. Uist, *J.W. Clark,* 1989 (BM).   16, 20, 21

**E. obscurum** Schreber
Short-fruited Willowherb.
Seileachan Fàireagach.
Native.

Roadsides.
Barra to Lewis.   1–6, 8, 10, 12, 16–19, 21

**E. palustre** L.
Marsh Willowherb.
Seileachan Lèana.
Native.
Marshes, flushes and streamsides.
Berneray and Mingulay to Lewis and St Kilda to Shiant Is.   1–7, 9–13, 15–24, 27–29
There is a record from Taransay of **E. palustre** × **parviflorum** by *J. W. H. Harrison,* but no specimen.

**E. × schmidtianum** Rostk.
**(obscurum × palustre)**
Benbecula *Shoolbred* 1894 (BM); Geshader, Uig, Lewis *Crabbe* 1939 (BM) det. *Ash.*   3, 18

**E. anagallidifolium** Lam.
Alpine Willowherb.
Seileachan Ailpeach.
Native.
There are specimens at BM from Uig, Lewis; Clisham, North Harris *Campbell* 1939 and Ben 'Caporl' (?Chaipaval) *Duncan* 1891. There is also a photograph of a living plant from above Loch Eynort, South Uist (BM). 3, 9, 12, ?19

**E. alsinifolium** Vill.
Chickweed Willowherb.
Seileachan Fliodhach.
Recorded from the Hecla/Beinn Mhor massif and Sheaval, South Uist by *H. H. Harrison* and *J. W. H. Harrison* but without specimens. ?19

**E. brunnescens** (Cockayne) Raven & Engelhorn
*E. nerteroides* auct.
New Zealand Willowherb.
Seileachan Làir.
Introduction.
Barra to Lewis; first record in Castlebay, Barra *Morton* 1957.   1, 3, 5, 7, 9, 18, 21
Scarce, perhaps only recently arrived and may be spreading.

**Chamerion** Rafin.
**C. angustifolium** (L.) J. Holub
*E. angustifolium* L.
Rosebay Willowherb.
Seileachan Frangach.
Colonist and possibly native.
Barra to Lewis.   1–3, 7, 8, 11, 16, 19, 21

**Fuchsia** L.
**F. magellanica** Lam. var. **macrostemma** (Ruiz & Pavón) Munz
Fuchsia.
Fiuise.**
Garden escape.
Tigha choddie, Northbay, Barra *Goodwin and Goodwin* 1951 (GL); field record from Eye penisula, 1959; North Uist, Lochmaddy, *J. W. Clark*, 1983.  6, 16, 21.

*HALORAGACEAE*
**Myriophyllum** L.
**M. spicatum** L.
Spiked Water-milfoil.
Snàthainn Bhàthaidh Spìceach.
Native.
Lochs.
Barra to Lewis.  2, 4, 11, 12, 15, 16, 18, 19, 21, 23
Prefers more base-rich water than the next species; often in water on machair.

**M. alterniflorum** DC.
Alternate Water-milfoil.
Snàthainn Bhàthaidh.
Native.
Streams and lochs.
Barra to Lewis.  1–7, 9–13, 16, 18–21, 29
Tolerates more oligotrophic conditions than the last species.

*HIPPURIDACEAE*
**Hippuris** L.
**H. vulgaris** L.
Mare's-tail.
Earball Capaill.
Native.
Lochs.
Barra to Lewis.  1, 7, 11, 12, 15, 16, 18, 19, 21, 29
Curiously absent from Mull and Rhum – very rare in Harris and Lewis.

*CALLITRICHACEAE*
**Callitriche** L.
**C. stagnalis** Scop.
Common Water-starwort.
Brailis Uisge, Biolair Ioc.
Native.
Ditches and disturbed boggy places.
Probably on most Islands from Berneray and Mingulay to Lewis and St Kilda to Shiant Is.  1–4, 6–12, 14–19, 21–24, 28, 29

**C. platycarpa** Kütz
Various-leaved Water-starwort.

Native.
There is BM material from Benbecula *Shoolbred* 1894; North Uist *Shoolbred* 1898; Scarp *W. S. Duncan* 1896 and Stornoway, Lewis *Somerville* 1900. There are literature records from several west coastal lochs in Lewis (Biagi et al. 1985) and St Kilda (Barrington, 1886).  1?, 2, 8, 16, 18, 28

**C. hamulata** Kütz. ex Koch
*C. intermedia* Hoffm.
Intermediate Water-starwort.
Native.
Vatersay to Lewis.  1–3, 6–8, 9, 11, 12, 16, 18, 19, 21, 22, 29

**C. hermaphroditica** L.
*C. autumnalis* L.
Autumnal Water-starwort.
Native.
Barra and Benbecula and the Uists.  16, 18, 19, 21

UMBELLIALES
*ARALIACEAE*
**Hedera** L.
**H. helix** L.
Ivy.
Eidheann.
Native.
Cliffs.
Scarce but widely distributed. Barra to Lewis.  3–5, 7, 10, 12, 16, 18, 19, 21, 24, 27

*UMBELLIFERAE*
**Hydrocotyle** L.
**H. vulgaris** L.
Marsh Pennywort.
Lus na Peighinn.
Native.
Damp machair, marshes and flushes.
Berneray and Mingulay to Lewis, St Kilda to North Rona and Shiant Is.  1–4, 6–9, 11–13, 15–25, 27–29

**Eryngium** L.
**E. maritimum** L.
Sea-holly.
Cuileann Tràgha.
Native.
On sands.
At present the Outer Hebrides records represent the most northerly localities for this species in the British Isles, since the Shetland plants have not been seen for over 50 years.
Mingulay to North Uist.  13, 16, 18, 19, 21, 23, 24

Chaerophyllum temulentum L.
Rough Chervil.
Only recorded for Benbecula and South Uist
(Harrison, 1941a). There are no specimens,
and this species is unlikely to occur.

**Anthriscus** Pers.
**A. sylvestris** (L.) Hoffm.
Cow Parsley.
Costag Fhiadhain.
Native.
Uncommon, grassy places.
Roadside, Castlebay, Barra *Edin. Univ. Biol.
Soc.* 1936 (E); there are literature records
from Mingulay and North Uist. 6, 12, 16,
17, 19–22, 24

**Myrrhis** Miller
**M. odorata** (L.) Scop.
Sweet Cicely.
Mirr, Còs Uisge.
Denizen.
Waste places near houses.
Lewis, Castle Park, Stornoway, *J. W.
Campbell* 1939 (BM), and Barvas, Harrison
(1957); North Uist, Newton, *M. S. Campbell*
1936 (BM), Balranald, *Meinertzhargen* 1947
(BM); South Uist, near houses at Dremisdale,
field record 1983, and on machair at Bornish,
Harrison (1957). 1/2, 5, 16, 19

**Smyrnium** L.
**S. olusatrum** L.
Alexanders.
Probably an escape.
Only record from Martin Martin in 1703
from North Uist where it was probably an
escaped pot-herb. It no longer occurs in the
Islands. See Perring (1953).

**Conopodium** Koch
**C. majus** (Gouan) Loret
Pignut.
Cnò-thalmhainn.
Possibly introduced.
Rare. South Uist, near Bornish House,
*Cadbury* 1978; Grimsay, field record 1983;
North Uist, Sponish Suspension Bridge,
Lochmaddy, *M. S. Campbell* 1937 (BM);
Lewis, Castle Park, Stornoway *M. S.
Campbell* 1938 (BM). 5, 16, 17, 19

**Aegopodium** L.
**A. podagraria** L.
Ground-elder.
Lus an Easbaig.
Colonist.

Cultivated ground.
Barra to Lewis.   1, 3, 6, 7, 9, 12, 14, 16, 18,
19, 21

**Berula** Koch
**B. erecta** (Hudson) Coville
Lesser Water-parsnip.
Folachdan.
Rare. Recorded from near Heisinish, Eriskay
(Harrison et al., 1939 and Harrison, 1941a);
South Uist, Loch na Liana Moire at Kilpheder,
*Cadbury* 1978.   19, 20

**Crithmum** L.
**C. maritimum** L.
Rock Samphire.
Saimbhir.
Native.
Cliffs.
Only certain record; north of Eilean Molach,
Ard More Mangersta, Uig, Lewis *Gibson*
1908 (BM); *M. S. Campbell* 1945 (BM). This is
the northern limit in Britain, Harrison et al.
(1939) also record it from cliffs south of West
Loch Tarbert, but there are no specimens.
Lewis.   3, [?12]

**Oenanthe** L.
**O. fistulosa** L.
Tubular Water-dropwort.
Native.
Extremely rare, known only from Kilpheder,
South Uist *M. S. Campbell* 1947 (BM); still
there in 1981 *Adams* (BM).

**O. lachenalii** C. C. Gmelin
Parsley Water-dropwort.
Dàtha Bàn.
Native.
Marshy pasture and loch margins.
South Uist, Benbecula, North Uist and
Monach Is.   16, 18, 19, 29

**O. crocata** L.
Hemlock Water-dropwort.
Dàtha Bàn Iteodha.
Native.
Streamsides.
Berneray to Harris.   7, 9–12, 16–21, 24

**Conium** L.
**C. maculatum** L.
Hemlock.
Iteodha.
Native or colonist.
Specimens seen from; Loch an Armainn,
North Uist *J. W. Campbell* 1939 (BM); by road

and sea at Castlebay, Barra *Sinclair* 1936 (E), and field record 1983, and a field record from Berneray in the Sound of Harris, 1962. There is a literature record from Barvas, Lewis *Balfour* and *Babington* (1844). [1/2], 15, 16, 21.

**Apium** L.
**A. nodiflorum** (L.) Lag.
Fool's Water-cress.
Biolair Brèige.**
Native.
Streams, wet ditches and wet marshes.
Vatersay to North Uist, Monach Is and Taransay.   11, 13, 15, 16, 18, 19, 21, 22, 29
Both this and the next species show a curious distibution, being absent from Mull and Skye but being not uncommon in the Outer Hebrides.

**A. inundatum** (L.) Reichenb. fil
Lesser Marshwort.
Native.
Fualactar.
Streams.
Barra to Lewis.   3, 4, 11–16, 18, 19, 21, 22, 29
See last species, although this one does occur on Iona.

**A.** × **moorei** (Syme) Druce
= A. inundatum × A. nodiflorum
Native.
Short turf by lochs.
Benbecula, Loch Mor, Edinburgh University Bot.Soc. 1951 (E); North Uist, Loch Coan on Baleshare, *McKean* 1984 (E).   16, 18.

**Cicuta** L.
**C. virosa** L.
Cowbane.
Fealladh Bog.
Very rare and restricted to South Uist, first record *Somerville* 1888 (BM), still at Loch Hallan in 1983 (RBG Edinburgh, 1983). However, in the elaboration of a lecture H. H. Harrison (1939) records it from 'Barra and/or Fuday' – the text is unclear.   19

**Carum** L.
**C. verticillatum** (L.) Koch
Whorled Caraway.
Probably native.
Only recorded from the Outer Hebrides by J. W. H.*Harrison* et al. Loch Eynort, South Uist 1939, 1945, 1947 (last two have specimens at K) and as in the last entry *H. H. Harrison*

records it indiscriminately from Barra and Fuday. Needs to be confirmed. It is interesting to note that on Mull it was not discovered until 1956.

**Ligusticum** L.
**L. scoticum** L.
Scots Lovage.
Sunais.
Native.
Maritime cliffs.
Berneray and Mingulay to Lewis and St Kilda, North Rona and Shiant Is.   1–4, 6, 7, 9–13, 16, 19, 21–25, 27–29
Okusanya (1979) reports on germination experiments on seeds of this species from Lewis, and remarks that it was better able to germinate in cold wet conditions than other species from similar habitats.

**Angelica** L.
**A. sylvestris** L.
Wild Angelica.
Lus nam Buadh, Aingealag.
Native.
Loch and streamsides.
Probably in all zones.   1–7, 9–25, 27, 28

**Heracleum** L.
**H. sphondylium** L.
Hogweed.
Odharan.
Native.
Machair and roadsides.
Berneray and Mingulay to Lewis.   1–4, 11–24, 29
The distinctive-looking dwarf plants on machair, with ternate instead of pinnate leaves, do not maintain these characters in cultivation.

**Torilis** Adanson
**T. japonica** (Houtt.) DC.
(*Caucalis anthriscus* Hudson)
Upright Hedge-parsley.
Peirsill Fàil.**
Doubtfully native.
Barra to Lewis.   1, 6, 9, 11, 16, 18, 21

**Daucus** L.
**D. carota** L. subsp. **carota**
Wild Carrot.
Curran Talmhainn.
Native.
Machair.
Berneray and Mingulay to Lewis.   1–4, 6, 7, 9, 11–13, 15, 16, 18–24, 27, 29

*EUPHORBIALES*
*EUPHORBIACEAE*
**Euphorbia** L.
**E. helioscopia** L.
Sun Spurge.
Lus nam Foinneachan.
Colonist.
Arable fields.
Mingulay to Lewis.   1, 3, 11, 12, 15, 16, 18, 19, 21, 22, 24, 29

**E. peplus** L.
Petty Spurge.
Lus Leighis.
Colonist.
Rare. Castlebay, Barra *Shoolbred* (1895), 'stray specimen'; amongst oats, Seilebost, South Harris, Harrison et al. (1941a).   12, 21.

POLYGONALES
*POLYGONACEAE*
**Polygonum** L.
**P. aviculare** L. sensu lato.
Knotgrass.
Glùineach Bheag.
Native.
Waste places and tracks.
Mingulay to Lewis and St Kilda.   1–3, 5–7, 9–12, 15–22, 24, 25, 28, 29

**P. arenastrum** Boreau
*P. aequale* Lindm.
Equal-leaved Knotgrass.
Native.
The only specimen which has been determined critically by *Styles* is from Barra, *Edin. Univ. Biol. Soc.* 1936 (E). There are literature records from Eriskay to Lewis, Harrison (1941a). There is also a record from St Kilda, village street, as *P. aequale Petch* (1933a).

**P. boreale** (Lange) Small
Northern Knotgrass.
Native.
Arable weed.
Rare. North Uist, *Balfour* 1841, two sheets at E det. by *Perring* and *Styles*; Eye, garden weed at Portvoller, *Noltie* 1987 det. *Akeroyd*, also at E.   6, 16

**P. oxyspermum** Meyer & Bunge ex Ledeb.
subsp. **raii** (Bab.) D. A. Webb & Chater
Ray's Knotgrass.
Glùineach na Tràighe.
Native.
Sandy seashores.

Pabbay (Barra) to Harris.   9, 12, 14, 18, 19, 21, 23

**P. viviparum** L.
Alpine Bistort.
Biolur Ailpeach.
Native.
Mountain rock ledges and grassy slopes.
South Uist to Lewis.   3, 4, 8, 9, 12, 19

**P. bistorta** L.
Common Bistort.
Biolur.
Hortal.
Only known from Lochboisdale, South Uist *Somerville* 1888 (BM), likely to have been a garden escape.

**P. amphibium** L.
Amphibious Bistort.
Glùineach an Uisge.
Native.
Lochs, marshes, machair, roadsides and waste places.
Barra to Lewis.   5, 11, 12, 15, 16, 18–22

**P. persicaria** L.
Redshank.
Glùineach Dhearg.
Doubtfully native.
Arable land, waste places and loch margins.
Mingulay to Lewis and St Kilda.   3–7, 9, 10, 12, 15–22, 24, 28, 29

**P. lapathifolium** L.
Pale Persicaria.
Possibly native.
Near Lochmaddy, North Uist *Shoolbred* 1894 (BM), the specimen is a poor one.   16

**P. hydropiper** L.
Water-pepper.
Glùineach Theth, Piobar Uisge.
Native.
South Harris, Obbe, *W. S. Duncan* 1891, 92 (BM), also literature records for Northton, Harrison (1954b); Killegray, ibid.; North Harris, Benbecula and Barra, Harrison (1941a). Apparently not seen recently.   9, 12, 14, 18, 21

**P. polystachyum** Wall. ex Meissner
Himalayan Knotweed.
Glùineach Hiomàilianach.
Hortal.
Waste ground near houses.

Near Tolsta, Lewis; Ardhasig, North Harris; both *Pankhurst* and *Chater* 1980 (BM).   1, 9

**Fallopia** Adanson
**F. convolvulus** (L.) A. Löve
(*Polygonum convolvulus* L.)
Black-bindweed.
Glùineach Dhubh.
Native or colonist.
Arable fields and waste places.
Barra to Lewis.   1, 3, 12, 15, 16, 18–22

**Reynoutria** Houtt.
**R. japonica** Houtt.
*Polygonum cuspidatum* Siebold & Zucc.
Japanese Knotweed.
Glùineach Sheapanach.
Naturalised.
Barra, Castlebay area, *Brookes*, 1978; South Uist, Stoneybridge; Benbecula *J. W. H. Harrison* (1941a); Ensay, *Morton*, 1958; South Harris, Harrison (1948); Lewis, North Dell, Harrison (1957).   1, 12, 14, 18, 19, 21

**R. sachalinensis** (F. Schmidt) Nakai
Giant Knotweed.
Glùineach Shagailìneach.
Introduced.
South Harris, along river bank at Borve Lodge, Harrison (1954a); Lewis, Stornoway Castle, Cunningham (1978).   5, 12

**Oxyria** Hill
**O. digyna** (L.) Hill
Mountain Sorrel.
Sealbhag nam Fiadh.
Native.
Mountain ledges.
Mingulay to Lewis and St Kilda. 3, 5, 9, 12, 18, 19, 23, 24, 28
Centred on the South Uist and North Harris massifs, surprisingly absent from Barra.

**Rumex** L.
**R. acetosella** L.
Sheep's Sorrel.
Sealbhag nan Caorach.
Native.
Dry banks, rocks and roadsides.
Berneray and Mingulay to Lewis and St Kilda.   1–7, 9–12, 15–22, 24, 25, 28

**R. acetosa** L.
Common Sorrel.
Sealbhag, Samh.
Native.
Marshes, machair and grassland.

Berneray and Mingulay to Lewis, and most of the outer islands.   1–7, 9–25, 27–29

**R. hibernicus** Rech. fil.
Irish Sorrel.
Native.
Taxonomic status unclear but two specimens determined by Rechinger; Atlantic coast of Barra *Cunningham* 1946 (K); Monach Is *Perring* 1949 (CGE).   21, 29

**R. longifolius** DC.
Northern Dock.
Copag Thuathach.
Native.
Surprisingly rare. Scalpay; South Harris, Luskentyre and Loch Stockinish, all *Wilmott* 1939 (BM); South Harris, Harrison (1941a); North Uist, under seawall at Lochmaddy, since destroyed, *J. W. Clark* 1981. 10, 12, 16, 21

**R. crispus** L.
Curled Dock.
Copag Chamagach.
Native.
Stony sea shores and waste places.
The var. **littoreus** Hardy, with shorter stems, dense inflorescences and nut 1.3–2.5 mm instead of 2.5–3.5 mm (Rich and Rich, 1988) is recorded from North Uist, and could be more frequent.
Probably in all zones.   1–9, 11–25, 27–29

**R. obtusifolius** L.
Broad-leaved Dock.
Copag Leathann.
Native.
Waste places.
Mingulay to Lewis, St Kilda and North Rona.   2–7, 9–12, 14–25, 27–29

**R. conglomeratus** Murray
Clustered Dock.
Copag Bhagaideach.
Recorded from St Kilda *Barrington* (1886) and Benbecula, North Uist and Harris *Shoolbred* (1895)
No specimens; status uncertain.   9 or 12, 16, 18

URTICALES
*URTICACEAE*
**Urtica** L.
**U. urens** L.
Small Nettle.
Deanntag Bhliadhnail, Feanntag Bhliadhnail.

Native.
Arable land and gardens.
Mingulay to Lewis.   1, 3, 6, 11, 12, 15, 16, 18–24

**U. dioica** L.
Common Nettle.
Deanntag, Feanntag.
Native.
Waste places.
Probably in all zones.   1, 3–7, 9–24, 27–29

**Humulus** L.
**H. lupulus** L.
Hop.
Lus an Leanna.
Introduction.
First seen in Glen Village on Barra *Wilmott* (1939); literature records by King's College (1939), Harrison (1941a).   21

*ULMACEAE*
**Ulmus** L.
**U. glabra** Hudson
Wych Elm.
Leamhan.
Planted.
Barra, southern end, field record 1988; North Uist, west of the Lodge, Newton *M. S. Campbell* 1936 (BM); Lewis, Castle Park, Stornoway, *M. S. Campbell* 1938 (BM) and Cunningham (1978).   5, 16, 21

MYRICALES
*MYRICACEAE*
**Myrica** L.
**M. gale** L.
Bog-myrtle.
Roid.
Native.
Damp heathy places.
Barra to Lewis; Campbell (1961) remarks on its patchy distribution and being commoner on the east of the Islands.   3, 5, 7, 9, 16, 18–21
*FAGALES*
*BETULACEAE*
**Betula** L.
**B. pendula** Roth
(*B. verrucosa* Ehrh., *B. alba* L.)
Silver Birch.
Beith Dhubhach.
Native or planted.
Stornoway Castle, Cunningham (1978) and field record 1980, probably planted; Rodel Glen, South Harris *Shoolbred* (1895) planted; Lochboisdale and Allt Volagir (not con-

firmed), South Uist, Harrison (1941a).   5, 12, 19

**B. pubescens** Ehrh.
Downy Birch.
Beith Charraigeach.
Native.
River ravines, sheltered cliffs.
Barra to Lewis.   1, 3, 5, 7, 9, 10, 12, 18, 19, 21
Bennett and Fossitt (1988) describe a B. pubescens woodland near Loch Eynort in South Uist, and mention another near Eishken.

**Alnus** Miller
**A. glutinosa** (L.) Gaertner
Alder.
Feàrna.
Doubtfully native.
Planted near houses.
There is a specimen from Allt Volagir, South Uist at BM which is presumably a native locality.
Barra to Lewis.   5, 7, 9, 12, 16, 19, 21

**A. incana** (L.) Moench
Grey Alder.
Planted, occasionally naturalised.
Lewis, Eishken Lodge, (Copping, 1976), grounds of Stornoway Castle, Harrison (1954a).   5, 7

*CORYLACEAE*
**Corylus** L.
**C. avellana** L.
Hazel.
Calltainn.
Native.
Mountain stream ravines, sea cliffs.
Barra to Lewis.   1, 3–5, 7–9, 12–14, 18–21

**Fagus** L.
**F. sylvatica** L.
Beech.
Faidhbhile.
Planted.
Stornoway Castle; Eishken; old plantation, Rodel Glen, South Harris *Shoolbred* (1895); Grogarry House, South Uist.
There is some palynological evidence that it used to occur on St Kilda and Lewis (Currie, 1979).   5, 12, 7, 19

**Castanea** Miller
**C. sativa** Miller
Sweet Chestnut.

Craobh Geanm-chnò.
Planted.
Stornoway Castle grounds, Cunningham
(1978); plantation, Northbay, Barra *Edin.
Univ. Biol. Soc.* 1936–7 (E), and still there
1983.   5, 21

**Quercus** L.
**Q. robur** L.
Pedunculate Oak.
Darach.
Planted.
Lewis, Stornoway Castle, Cunningham
(1978); South Harris, old plantation, Rodel
Glen, (Shoolbred, 1895).   5, 12

**Q. petraea** (Mattuschka) Liebl.
*Q. sessiliflora* Salisb.
Sessile Oak.
Darach.
Planted.
Castle Park, Stornoway, Lewis *M. S. Campbell*
1938 (BM).   5

SALICALES
*SALICACEAE*
**Populus** L.
**P. tremula** L.
Aspen.
Critheann.
Native.
On cliffs. Mingulay to Lewis.   2, 4, 5, 7, 12,
16, 18–21, 24

**Salix** L.
**S. pentandra** L.
Bay Willow.
Planted.
Barra to South Harris.   12, 16, 21

**S. fragilis** L.
Crack Willow.
Planted.
Rodel, South Harris *Balfour & Babington*
(1844); Benbecula and South Uist (Harrison,
1941a).   12, 18, 19

**S. alba** L.
White Willow.
Planted.
Glen of Rodel, South Harris *Balfour &
Babington* (1844); Tarbert, Harris *Shoolbred*
(1895); roadside west of Grogarry House,
South Uist *Chater* 1983 (BM).   12, 19

**S. purpurea** L.
Purple Willow.

Planted.
Along the Lochs road, Stornoway, Harrison
(1957); near the house at Eishken, *Chater &
Pankhurst* 1980.   5, 7

**S. viminalis** L.
Osier.
Seileach uisge.
Planted.
Gardens.
Barra to Lewis.   3, 4, 7, 9, 12, 15–19, 21

**S.** × **rubra** Hudson.
= S. purpurea L. × S. viminalis L.
Lews Park, Stornoway, Lewis (Copping,
1977).   5

**S.** × **calodendron** Wimm.
see Meikle, 1952.
Woods around Borve Lodge, South Harris
*M. S. Campbell* 1938 (BM) det. (as **S. dasy-
clados** Wimm.) *Neumann* 1961

**S. caprea** L.
Goat Willow.
Geal-sheileach, Sùileag.
Planted and perhaps native.
Planted in Castle Park, Stornoway, Lewis
*Wilmott* 1946 (BM); by stream at head of
Bagh Huilavagh, Barra *Chater* 1983 (BM). 5,
21

**S. cinerea** L.
Grey Willow.
Dubh-sheileach.
Native.
Loch margins and islands.
subsp. **oleifolia** Macreight
= S. atrocinerea Brot.
Barra to Lewis.   2, 4, 5, 7, 9, 12, 16, 18, 19, 21
**S.** × **smithiana** Willd.
= S. cinerea L. × S. viminalis L.
By Lodge at Eishken, Lewis *Pankhurst &
Chater* 1980 (BM) det. *Meikle*.   7

**S.** × **multinervis** Doell.
= S. aurita L. × cinerea L.
Northbay plantation, Barra *Edin. Univ. Biol.
Soc.* 1936 (E) det. *Wilmott*; Also recorded
from North and South Uist and Benbecula
(Harrison 1941a, Harrison et al. 1942a)

**S.** × **subsericea** (S. cinerea L. × S. repens L.)
has been recorded from Corrie Dubh, South
Uist and from north-east Barra (Harrison,
1941a).
There are no specimens.

**S. aurita** L.
Eared Willow.
Seileach Cluasach.
Native.
Loch margins, streamsides and moorland.
Berneray and Mingulay to Lewis.    1, 3–5,
7–10, 12, 13, 16–24, 27

**S. × coriacea** Forbes
= S. aurita × S. myrsinifolia Salisb.[nigricans
Sm.]
Glen Tealasdale, Uig, Lewis *Wilmott* and
*M. S. Campbell* 1939 (BM)

**S. × ambigua** Ehrh.
= S. aurita L. × S. repens L.
Mingulay to Lewis.    3, 4, 9, 12, 16, 19, 24

**S. myrsinifolia** Salisb.
*S. nigricans* Sm.
Dark-leaved Willow
Seileach na Duilleige Duirche.
Native.
River banks, river ravines and cliffs.
South Uist to Lewis.    3, 5, 7, 9, 12, 19
Var. **hebridensis** Wilmott is described in
detail by Wilmott in the Flora of Uig
(Campbell, 1945), p.50 from Glen Tealasdale.
It is described as having oblanceolate leaves,
to 70(–90) by 25 cm (sic) which are conspicu-
ously and sharply crenate-dentate, and
scarcely turning black on drying.    3

**S. phylicifolia** L.
Tea-leaved Willow.
Native.
River banks, river ravines and cliffs.
South Uist to Lewis.    3, 7, 9, 19

**S. repens** L.
Creeping Willow.
Seileach Làir.
Native.
Machair and moorland.
For an account of the variation in this
species, see Meikle (1984)

subsp. **repens**
Berneray and Mingulay to Lewis, St Kilda
and Shiant Is.    1–22, 24, 27–29

subsp. **argentea** (Sm.) A. & G. Camus
inc. S. arenaria L.
Barra to Lewis.    1, 3, 6, 9, 15, 16, 18, 19, 21,
23
The records of these taxa appear much confused; the second subspecies is erect and
normally (but not always) dune-dwelling.

**S. × pseudospuria** Rouy (S. arbuscula L. ×
lapponum L.) has been recorded from a
ledge in Glen Skeaudale, North Harris
(Harrison, 1941a). *J. W. H. Harrison* says
'determined by Kew as this hybrid' but the
specimen has not been traced, and he main-
tains that he considers the specimen to be a
Hebridean form of S. lapponum L.

**S. herbacea** L.
Dwarf Willow.
Seileach Ailpeach.
Native.
High moorland.
Barra to Lewis.    3, 9, 12, 16, 19, 21, 28
See comment under *Hymenophyllum wilsoni*.
A specimen from St Kilda *Gibson* 1889 (E)
was determined as **S. × cernua** E. F. Linton
(S. herbacea L. × S. repens L.) by *Linton* but
is not included by Meikle (1984).

ERICALES
*ERICACEAE*
**Rhododendron** L.
**R. ponticum** L.
Rhododendron.
Rodaideandran.
Planted.
Stornoway Castle; by house at Eishken;
on North Harris estate; North side of L.
Druidibeg on road to Lochskiport.    5, 7, 9,
19

**Loiseleuria** Desv.
**L. procumbens** (L.) Desv.
Trailing Azalea.
Lusan Albannach.
One literature record only, from Clisham
(Harris), unconfirmed by voucher material,
Gordon (1923).    9

**Arctostaphylos** Adanson
**A. uva-ursi** (L.) Sprengel
Bearberry.
Grainnseag, Braoileag nan Con.
Native.
Cliffs.
South Uist to Lewis.    3–5, 9, 12, 15, 16, 19

**Vaccinium** L.
**V. vitis-idaea** L.
Cowberry.
Lus nam Braoileag.
Native.
Recorded from the high ground in North
Harris and near Stornoway and also from St
Kilda, surprisingly uncommon. Harrison

(1941a) records it from Hecla, South Uist.
?5, 9, 19, 28

**V. myrtillus** L.
Bilberry.
Caora-mhitheag.
Native.
Moorland.
Barra to Lewis and St Kilda.    3, 5, 7, 9–13,
16–21, 27, 28

**Calluna** Salisb.
**C. vulgaris** (L.) Hull
Heather.
Fraoch.
Native.
Heathland and moorland.
Probably in all zones.   1–7, 9–24, 27–29

**Erica** L.
**E. tetralix** L.
Cross-leaved Heath.
Fraoch Frangach, Fraoch Gucagach.
Native.
Damp moorland and bogs.
Berneray and Mingulay to Lewis, St Kilda
and Shiant Is.    1, 3–7, 9–13, 16–24, 27, 28

**E. cinerea** L.
Bell Heather.
Fraoch a' Bhadain, Biadh na Circe-fraoich.
Native.
Dry moorland and rocks.
Mingulay to Lewis and St Kilda.    1–7, 9–13,
15–24, 27, 28

*PYROLACEAE*
**Moneses** Salisb.
**M. uniflora** (L.) A. Gray
One-flowered Wintergreen.
Possibly native.
There are two specimens in Herb. Smith at
LINN labelled 'From the western Isles of
Scotland Harris and Bernera gathered in
1783 by Jas.Hoggan'. These were communi-
cated to Smith in December 1793 by *R.
Gotobed*. The record is published in the
Annals Scot. Nat. Hist. 1911:185. The locali-
ties given are highly ambiguous, since
Harris is a large area and Bernera might refer
to Great or Little Bernera, the island of
Berneray in the Sound of Harris, or even the
island of Berneray to the south of Barra. No
information is given about the habitat, but
Moneses is a plant of montane, usually
coniferous, woodland. This type of vegetation is believed to have been absent from

the Outer Hebrides within the period of
recorded history. *M. S. Campbell* was of the
opinion that the specimens had been mis-
labelled. If so, they might have come from
eastern Scotland. If Moneses ever did grow
in the Outer Hebrides, it is virtually certain
now to be extinct.
This is a protected species, listed in the Red
Data Book, Perring and Farrell (1983).

*EMPETRACEAE*
**Empetrum** L.
**E. nigrum** L.
Crowberry.
Lus na Feannaig.
Native.
Moorland and sea cliffs.
subsp. **nigrum**
Barra to Lewis, St Kilda to Shiant Isles.  1,
3, 4, 9, 11–13, 16, 18, 19, 21, 22, 24, 27, 28
All the BM specimens of this have been
examined without our being able to dis-
tinguish the next subspecies; many however
are sterile and not identifiable to subspecies.
The same difficulty is remarked upon by
Scott and Palmer (1987) for Shetland, and
yet apparently the plant fruits freely in the
Faeroes.

subsp. **hermaphroditum** (Hagerup) Böcher
*E. hermaphroditum* Hagerup
Mountain Crowberry.
Dearcag Fithich.
Native.
Literature records only; islet in loch south of
Loch Uamavat, Harris? (Harrison, 1956);
saddle between Clisham and Tomnaval,
North Harris (Harrison et al., 1941a).
The easiest way to distinguish this sub-
species is to examine the fruit for the shriv-
elled remains of stamens, which proves that
its flowers were perfect. It is normally a
plant of rocks at high altitude.

PLUMBAGINALES
*PLUMBAGINACEAE*
**Armeria** Willd.
**A. maritima** (Miller) Willd.
Thrift.
Neòinean Cladaich, Tonn a' Chladaich.
Native.
Coastal rocks, salt marshes, moorlands and
mountains.
Common throughout the Islands.
Probably in all zones.   1–25, 27–29

PRIMULALES
*PRIMULACEAE*
**Primula** L.
**P. vulgaris** Hudson
Primrose.
Sòbhrach.
Native.
Streamsides and rocky banks.
Berneray and Mingulay to Lewis, St Kilda and Shiant Is.   2–7, 9–24, 27, 28

**Lysimachia** L.
**L. nemorum** L.
Yellow Pimpernel.
Seamrag Moire.
Native.
Rare; field record at the head of Loch Shell, *Chater & Pankhurst* 1980; there are specimens from Obbe, South Harris *Duncan* 1891 (BM); Rodel, South Harris *M. S. Campbell* 1939 (BM); East side Beinn Mhor, South Uist *Warburg* 1947 (BM); Barra from Northbay wood, *Edin. Univ. Biol. Soc.* 1937 (E), and field record, Ersary; also some literature records from these islands by *W. A. Clark* and *J. W. H. Harrison.*   7, 12, 19, 21, 22

**Anagallis** L.
**A. tenella** (L.) L.
Bog Pimpernel.
Falcair Lèana.
Native.
Flushes and damp machair.
Berneray and Mingulay to Lewis, St Kilda and Monach Is.1–4, 6, 9, 11–24, 27–29

**A. arvensis** L.
Scarlet Pimpernel.
Falcair.
Possibly native.
Arable fields and gardens.
Mingulay to Barra and Fuday, and the Uists. 16, 18, 19, 21–24

**A. minima** (L.) E. H. L. Krause
*Centunculus minimus* L.
Chaffweed.
Falcair Mìn.
Native.
Damp spots and *Festuca rubra* zone of saltmarshes.
Barra to Lewis.   2, 3, 8, 9, 11, 12, 15–21, 24

**Glaux** L.
**G. maritima** L.
Sea-milkwort.
Lus Bàn-dhearg a' Chladaich.*

Native.
Salt marshes, lochsides near the sea and maritime rocks.
Probably in all zones.   1–5, 7–10, 12–25, 27, 29

**Samolus** L.
**S. valerandi** L.
Brookweed.
Lus Beag Bàn an Uillt.*
Native.
Marshes.
Barra to Lewis and Monach Is.   3, 12, 15, 16, 18, 19, 21, 29

CONTORTAE
*OLEACEAE*
**Fraxinus** L.
**F. excelsior** L.
Ash.
Uinnseann.
Native or planted.
First recorded at Rodel, South Harris by *Balfour & Babington* in 1841. Literature records from woods at Luskentyre and Stornoway. Currie (1979) gives a record for a sapling seen in Allt Volagir by *J. Balfour*. Tree by road at head of Bagh Huilavagh, Barra *M. Cannon* 1983 (BM).   5, 7, 12, 19, 21

**Ligustrum** L.
**L. vulgare** L.
Wild Privet.
Ras-chrann Sìor-uaine.
Planted.
Garden hedges.
Field records only. South Uist, near Loch Druidibeg, *Ratcliffe* 1960; Lewis, Arivruaich, *Fox,* 1959.   5/7, 19

*GENTIANACEAE*
Cicendia filiformis (L.) Delarb. has been recorded from Fuday and from the sandy shore of a loch in South Uist (Harrison, 1941a, b). There appear to be no specimens and the plant is otherwise unknown from the Islands and unlikely to occur.

**Centaurium** Hill
**C. erythraea** Rafn
Common Centaury.
Ceud-bhileach, Deagha Dearg.
Native.
Machair.
Berneray and Mingulay to Lewis and St Kilda.   3, 9, 11, 12, 13, 15–24, 28

**Gentianella** Moench
**G. campestris** (L.) Börner
Field Gentian.
Lus a' Chrùbain.
Native.
Machair, roadside banks and dry grassland.
Sandray to Lewis, St Kilda, Monach Is,
Shiant Is. 1–4, 6, 7, 9, 11–13, 15–24, 27–29

**G. amarella** (L.) Börner
subsp. **septentrionalis** (Druce) Pritchard
Northern Autumn Gentian.
Native.
This plant is strikingly different from southern forms of the species in having a purple
corolla with a yellow throat, see Pritchard
(1950). BM specimens come from the NE
coast of Lewis between Stornoway and the
Butt of Lewis. There is a population of
depauperate plants at Berie sands, Uig
*Pankhurst*, 1980. In addition there are literature records from near Valtos (probably
Berie), and between Borve and Scarasta,
South Harris (Harrison et al. 1941a). 1, 3,
12

*MENYANTHACEAE*
**Menyanthes** L.
**M. trifoliata** L.
Bogbean.
Trì-bhileach, Pònair Chapaill.
Native.
Shallow lochs and wet bogs.
Throughout the islands. 1–7, 9–13, 15–23,
27, 29

TUBIFLORAE
*BORAGINACEAE*
**Symphytum** L.
**S. officinale** L.
Common Comfrey.
Meacan Dubh.
Probably hortal.
Field record at Rodel, South Harris; recorded
from Newton Lodge, North Uist as var.
**patens** Sibth. by *J. W. Campbell* and *Wilmott*
(Campbell, 1937b); field record at Balvanich,
Benbecula, 1979. 12, 16, 18

**S. × uplandicum** Nyman
(*S. asperum* Lepechin × *S. officinale* L.)
Russian Comfrey.
Meacan Dubh Ruiseanach.
Hortal.
Grimsay, Baymore, *Chater* 1983 (BM); Lewis,
Port of Ness, *Cadbury* 1978; there is also a
specimen from Ensay, Sound of Harris *M. S.*

*Campbell* 1939 (BM) but this is just a leaf and
not identifiable for certainty. 1, 14?, 17
Symphytum tuberosum L. has been
recorded from dunes at Scarastavore by
Harrison et al. (1941a).

**Pentaglottis** Tausch
**P. sempervirens** (L.) Tausch
Green Alkanet.
Bog-lus.
Introduction.
Field records only. From square NF81 (South
Uist or Eriskay) and from NF86 (North Uist),
both about 1950. Requires confirmation. 16,
19/20

**Anchusa** L.
**A. arvensis** (L.) Bieb.
(*Lycopsis arvensis* L.)
Bugloss.
Lus Teanga an Daimh.
Arable weed.
Mingulay to Lewis. 1, 3, 11–13, 15, 16, 18,
19, 21–24, 29

**Nonea** Medic.
**N. rosea** (Bieb) Link
Casual in oats, Baleloch, North Uist *Shoolbred*
(1898).

**Myosotis** L.
**M. scorpioides** L.
(*M. palustris* (L.) Hill)
Water Forget-me-not.
Cotharach.
Native.
Rare and scattered.
Vatersay, Barra, North Uist, Lewis, Monach
Is and Shiant Is. 1, 3, 16, 21, 22, 27, 29

**M. secunda** A. Murray
Creeping Forget-me-not.
Lus Midhe Ealaidheach.
Native.
Damp places and ditches.
Barra to Lewis and Shiant Is. 1, 4, 12, 14,
16, 18, 19, 21, 22, 27

**M. laxa** Lehm.
subsp. **caespitosa** (C. F. Schultz) Hylander
ex Nordh.
*M. caespitosa* C. F. Schultz
Tufted Forget-me-not.
Lus Midhe Dosach.
Native.
Marshes, streamsides and flushes.
Mingulay to Lewis. 1–4, 6, 7, 9–19, 21–24, 29

**M. arvensis** (L.) Hill
Common Forget-me-not.
Lus Midhe Aitich.
Native.
Arable land.
Mingulay to Lewis and St Kilda but appar
ently absent from North Harris.  1–4, 10–19,
21–24, 28, 29

**M. discolor** Pers.
Changing Forget-me-not.
Lus Midhe Caochlaideach.
Native.
Machair and bare places.
Mingulay to Lewis and Shiant Isles.  1–5, 6,
8–10, 12, 16, 18, 19, 21–24, 27

**Lithospermum** L.
**L. arvense** L.
Field Gromwell.
Introduced.
Cultivated fields.
Eriskay to Lewis.  3, 11, 16, 19, 20

**Mertensia** Roth
**M. maritima** (L.) Gray
Oysterplant.
Tiodhlac na Mara.*
Native.
Rare. May have declined due to grazing by
sheep (Randall, 1979).
*Balfour* and *Babington* (1844) recorded it from
near Stornoway, Lewis in 1841. Reported
from Baleshare, North Uist by *Murray*.
Recorded on shingle beach at Nunton,
Benbecula, in 1979 by *Pankhurst et al.*, one
large plant and several seedlings, photo at
BM. The only record seen backed with a
specimen is: small shingle beach, Stockay
(Monach Is.) *Perring* 1949 (CGE). There is a
photograph of this in Ranwell (1977).  16,
18, 29

**Echium** L.
**E. vulgare** L.
Viper's-bugloss.
Lus na Nathrach.
Introduced.
Machair.
Taransay *Shoolbred* 1894 (BM) confirmed
*Angus* 1989, also literature record by *Clark*
(1939c), near Scarista, South Harris *Balfour
& Babington* 1841.  11, ?12

*CONVOLVULACEAE*
**Convolvulus** L.
**C. arvensis** L.

Field Bindweed.
Iadh-lus.
Doubtfully native.
Arable land.
East of ruined church, Risgary, Berneray
(Harris) *Chorley* and *Pankhurst* 1983 (BM); in
rye near village, Vatersay (Harrison, 1941a).
15, 22

**Calystegia** R. Br.
**C. sepium** (L.) R. Br.
subsp. **sepium**
Hedge Bindweed.
Dùil Mhial.
Introduction.
Barra, South Uist, South Harris and Lewis.
2, 6, 12, 19, 21

subsp. **silvatica** (Kit.) Griseb.
*C. silvatica* (Kit.) Griseb.
Large Bindweed.
Dùil Mhial Mhòr.
Introduction.
Field records from South Harris, Rodel,
*Pankhurst* 1979 and Renish Point, *Dony*, 1959;
Eishken House, *Pankhurst & Chater* 1980.
Literature record from coast near
Leverburgh, Harrison (1948).  7, 12

**C. soldanella** (L.) R. Br.
Sea Bindweed.
Flùr a' Phrionnsa.
Native.
*Ammophila* dunes.
This plant is alleged to have been sown on
Eriskay by Prince Charles when he arrived
in 1745.
Eriskay and Vatersay.  20, 22

*SCROPHULARIACEAE*
**Linaria** Miller
**L. vulgaris** Miller
Common Toadflax.
Buabh-lion Coitcheann.**
Probably introduced.
Rare, field record only. Lewis, cultivated
ground at Callanish, *Roe*, 1959.  2

**Cymbalaria** Hill
**C. muralis** P. Gaertner, B. Meyer & Scherb.
Ivy-leaved Toadflax.
Buabh-lion Eidheannach.**
Introduction.
Old greenhouse, Castle Park, Stornoway,
Lewis *J. W. Campbell* 1949 (BM).  5

**Scrophularia** L.
**S. nodosa** L.

Common Figwort.
Lus nan Cnapan.
Possibly native.
Below the garden, Newton, North Uist *M. S. Campbell* 1936 (BM); cliffs, south shore of Loch Eynort, South Uist (Harrison, 1941a). 16, 19

**S. auriculata** L.
*S. aquatica* auct. non L.
Water Figwort.
Lus nan Cnapan Uisge.
Possibly introduced.
Lewis, Stornoway Woods, Harrison (1957). Apparently not seen since and requires confirmation.   5

**Mimulus** L.
**M. guttatus** DC.
Monkeyflower.
Meilleag an Uillt.
Introduction.
First record Castle Park, Stornoway, Lewis *M. S. Campbell* 1938 (BM), det. *Silverside*; reported from a stream at Tarbert, North Harris (Harrison et al., 1941a); Grogarry, South Uist *M. S. Campbell* et al. (BM).   5, 9, 19

**M. guttatus** DC. × **luteus** agg.
Material of this parentage has been determined by *Silverside* in two groups;

**M. guttatus** × **luteus** L. s.s.
North of Loch Odhairn, Gravir, Lewis *Pankhurst* and *Chater* 1980 (BM) only record. 7

**M. × caledonicus** Silverside
? = *guttatus* × *luteus* × *variegatus*
South Harris, Manish, *Pankhurst* and *Chater* 1980 (BM); Lewis, Barvas, *M. S. Campbell* 1937 and 1938 (BM), south of Gress River, Stornoway, *M. S. Campbell* 1938 (BM), L. Stiapavat, Biagi et al. (1985), New Tolsta, *Noltie*, 1987.   1, 12

M. luteus L.
Blood-drop-emlets.
According to Silverside, field records of this species in the 1950s from North Uist and Lewis, zones 1, 2, 5 and 16, are probably referable to one of the above hybrids.

**Sibthorpia** L.
**S. europaea** L.
Cornish Moneywort.

Although only native in the extreme SW of Britain and Ireland this plant has been known as an introduction in the grounds of Castle Park, Stornoway, Lewis for over forty years; the earliest record appears to be *Crabbe, M. S. Campbell* and *Wilmott* 1946 (BM).   5

**Digitalis** L.
**D. purpurea** L.
Foxglove.
Lus nam Ban-sìdh.
Native.
Sheltered banks, woods.
South Uist to North Harris.   5, 7, 9, 12, 16, 18, 19

**Veronica** L.
**V. beccabunga** L.
Brooklime.
Lochal Mothair.
Probably native.
Rare in the Islands, first recorded from Barra *Shoolbred* (1895), still at Castlebay, 1983; south of Gress River, Gress, Lewis *M. S. Campbell* 1938 (BM); marsh near Sollas aerodrome, North Uist *Clark* (1939c), and at Balranald, *Murray* 1978; Harrison (1941a) also records it from Barra, Fuday and Vatersay, and from South Harris (Harrison, 1957).   1, 12, 15, 16, 21, 22

**V. anagallis-aquatica** L.
Blue Water-speedwell.
Fualachdar.
Native.
Streams and wet ditches.
Vatersay to Lewis and Monach Is.   1, 3, 12–16, 18, 19, 21, 22, 29

V. × lackchewitzii Keller
= *V. anagallia-aquatica* L. × *V. catenata* Pennell.
Bay East of Linique, South Uist *M. S. Campbell* and *Warburg* 1947 (BM) although available this determination was not confirmed by J. H. Burnett but the specimen appears correctly named.   19

**V. catenata** Pennell
Pink Water-speedwell.
Lus-crè Uisge.
Native.
South Uist to Berneray, also Butt of Lewis. 1, 13, 15, 16, 18, 19

**V. scutellata** L.
Marsh Speedwell.

Lus-crè Lèana.
Native.
Marshes.
The var. **villosa** Schum., with stems densely long hairy, might occur but there are no records.
South Uist to Lewis.    2–5, 7, 12, 16, 18, 19, 29

**V. officinalis** L.
Heath Speedwell.
Lus-crè Monaidh.
Native.
Dry banks and rocks.
Barra to Lewis and St Kilda.    2, 3, 5, 7, 9, 12, 13, 16, 18, 19, 21, 22, 28

**V. chamaedrys** L.
Germander Speedwell.
Nuallach.
Possibly introduced.
Records are all from inhabited areas; Stornoway, Lochmaddy and Castlebay. Commonest on North Uist.
Barra to Lewis.    5, 16, 21

**V. serpyllifolia** L.
Thyme-leaved Speedwell.
Lus-crè Talmhainn.
Native.
Roadsides.
Mingulay to Lewis.    1, 3–7, 9, 12, 16–19, 21, 24

**V. arvensis** L.
Wall Speedwell.
Lus-crè Balla.
Probably native.
Machair.
Mingulay to Lewis.    1–7, 9, 11–13, 15, 16–18, 19, 21–24, 29

**V. persica** Poiret
Common Field-speedwell.
Lus-crè Gàrraidh.
Introduction.
Arable Land.
Scarce. First record Rodel Lodge, South Harris *Shoolbred* (1895); field records from Barra, 1988; Grimsay, 1983; Askernish, South Uist *M. S. Campbell* et al. 1947 (BM); Borve, South Harris *Park* 1951 (BM). There are also literature records from: Keepers House, Mingulay *Clark* (1938); Tarbert, North Harris *Druce* (1929); near Lochboisdale, South Uist *J. W. H. Harrison* et al., (1942b).    9, 12, 17, 19, 21, 24

**V. agrestis** L.
Green Field-speedwell.
Lus-crè Arbhair.
Colonist.
First record North Uist *Shoolbred* (1895); Peinylodden, Benbecula *M. S. Campbell* and *Wilmott* 1947 (BM); Taransay, Harris *Poore* (BM); *J. W. H. Harrison* also recorded it from South Uist and Lewis.    3, 11, 16, 18, 19, ?21, ?22

**V. filiformis** Sm.
Slender Speedwell.
Lus-crè Claidh.
Recent introduction.
The only record: garden, Lochmaddy, North Uist *J. W. Clark*, 1980.    16

**Pedicularis** L.
**P. palustris** L.
Marsh Lousewort, Red-rattle.
Lus Riabhach.
Native.
Marshes, flushes and lochsides.
Barra to Lewis.    1–7, 9–23, 29

**P. sylvatica** L.
subsp. **sylvatica**
Lousewort.
Lus Riabhach Monaidh.
Native.
Moorland.
Berneray and Mingulay to Lewis, Shiant Is and St Kilda    2–7, 9–24, 27, 28

Subsp. **hibernica** D. A. Webb
Distinguished by having the calyx and pedicels with long, white curled hairs instead of being glabrous. The first record from outside Ireland was from St.Kilda.
Scalpay, Harris *J. W. Campbell*; also *Wilmott, M. S. Campbell* and *Bangerter* 1939 (BM); St.Kilda, *Gladstone* 1927 (K).    10, 28

**Rhinanthus** L.
**R. minor** L.
Yellow-rattle.
Modhalan Buidhe.
Native.
Machair and roadsides.
On machairs and roadsides throughout the Islands.    1–4, 6, 7, 9–13, 15–24, 27–29
The treatment of this genus is confused and few of the specimens available have been named to subspecies and are not easy to identify as dried material. See the key in Rich and Rich (1988). The subspecies

recorded are: subspp. **minor, stenophyllus** (Schur) O. Schwarz, **monticola** (Sterneck) O. Schwarz (= *R. spadiceus* Wilmott) and **borealis** (Sterneck) P. D. Sell.

**Melampyrum** L.
**M. pratense** L.
Common Cow-wheat.
Càraid Bhuidhe.*
Native.
Very rare, no specimens seen. Near the Obbe, South Harris *Balfour* and *Babington* 1841; Clisham, North Harris *Shoolbred* (1895)

and *J. W. H. Harrison* (1941a); Islet in Loch Drinishader, South Harris *J. W. H. Harrison* (1956); Barra *Conacher* (1980).   9, 12, 21

**Euphrasia** L.
(**E. officinalis** L. agg.)
Eyebright.
Lus nan Leac.
Native.
Machair, maritime heath, moorland and mountain ledges.
The first record (as the aggregate *E. officinalis*) was made by Balfour and Babington in 1841.   1–25, 27–29

The species of this genus are difficult taxonomically since inbreeding and hybridisation have given rise to locally distinct populations. Valid species with wide distributions are recognisable, but these tend to be rather similar to one another and in some cases the range of morphological variation of one species can overlap with another. Fertile hybrids occur frequently and it is often impossible to name them with certainty. It is usually unwise to attempt to name a population from which only a single plant has been collected. Dried specimens lose all their colour and the stems and leaves turn black. The length of the corolla is an important taxonomic character, but the corolla shrinks somewhat after drying. The significant characters of leaf shape, indumentum and toothing become difficult to see after drying as the leaves fold and cover one another. Some species may have the flowers in four distinct rows (**quadrate**), but this is impossible to see in dried material. This account relies heavily on specimens determined by P. F. Yeo and on his monograph of the European species (Yeo, 1978). Help and advice from A. J. Silverside is also acknowledged. In the following key, the species definitions have been treated *sensu stricto*, since otherwise the species overlap and it is impossible to write a definitive key. Two of the most important contributors to our knowledge of the flora of the Outer Hebrides, *Miss M. S. Campbell* and *Prof. J. W. Heslop Harrison* are honoured in this genus.

### Key to *Euphrasia* in the Outer Hebrides

1 Corolla length (measured from the base of the tube to the tip of the upper lip) more than 7.5 mm . . . . .2
Corolla length not more than 7.5 mm . . . . . . . . . . . . . . . . . . . . . . . . . . . . . . . . . . . . . . . . . . . . . . . .5

2 Basal pairs of teeth of lower floral leaves pointing forwards (antrorse) . . . . . . . . . . . . . . . . . . . . . *confusa*
Basal pairs of teeth of lower floral leaves patent . . . . . . . . . . . . . . . . . . . . . . . . . . . . . . . . . . . . . . . .3

3 Lowest flower at node 8 or lower (count nodes from the base but do not include the cotyledonary
pair) . . . . . . . . . . . . . . . . . . . . . . . . . . . . . . . . . . . . . . . . . . . . . . . . . . . *arctica* subsp. *borealis*
Lowest flower at node 9 or higher . . . . . . . . . . . . . . . . . . . . . . . . . . . . . . . . . . . . . . . . . . . . . . . .4

4 Stem and branches usually slender and flexuous, leaves near base of branches usually very
small . . . . . . . . . . . . . . . . . . . . . . . . . . . . . . . . . . . . . . . . . . . . . . . . . . . . . . . . . . . *confusa*
Stem and branches usually stout and not flexuous, leaves near base of branches not noticeably
smaller than the others . . . . . . . . . . . . . . . . . . . . . . . . . . . . . . . . . . . . . . . . . . . . . *nemorosa*

5 Calyx-tube whitish and membranous, with prominent green veins . . . . . . . . . . . . . . . . . . . *campbelliae*
Calyx-tube green and herbaceous . . . . . . . . . . . . . . . . . . . . . . . . . . . . . . . . . . . . . . . . . . . . . . . . .6

6 Lowest flower at node 6 or higher . . . . . . . . . . . . . . . . . . . . . . . . . . . . . . . . . . . . . . . . . . . . . . . . . .7
Lowest flower at node 5 or lower  . . . . . . . . . . . . . . . . . . . . . . . . . . . . . . . . . . . . . . . . . . . . . . . . .23

7 Cauline internodes usually 2–6 times as long as leaves . . . . . . . . . . . . . . . . . . . . . . . . . . . . . . . . . .8
Cauline internodes mostly not more than twice the leaves  . . . . . . . . . . . . . . . . . . . . . . . . . . . . . . .17

8 Basal pairs of teeth of lower floral leaves pointing forwards . . . . . . . . . . . . . . . . . . . . . . . . . . . . . . .9
Basal pairs of teeth of lower floral leaves patent . . . . . . . . . . . . . . . . . . . . . . . . . . . . . . . . . . . . . . .11

9 Teeth of lower floral leaves obtuse to acute, fairly short, corolla not more than 6.5 mm . . . . . . . . . . . .10
Teeth of lower floral leaves acute to aristate, corolla 7mm or more . . . . . . . . . . . . *arctica* subsp. *borealis*

10  Leaves strongly tinged with purple, not darker beneath than above, corolla usually lilac to purple, capsule shorter than calyx . . . . . . . . . . . . . . . . . . . . . . . . . . . . . . . . . . . . . . . . . . . . . **micrantha**
    Leaves faintly or moderately purplish, often darker below than above, corolla usually white, capsule at least as long as calyx . . . . . . . . . . . . . . . . . . . . . . . . . . . . . . . . . . . . . . . . . . . . **scottica**

11  Corolla length 6.5 mm or more . . . . . . . . . . . . . . . . . . . . . . . . . . . . . . . . . . . . . . . . . . . . . . . . . 12
    Corolla less than 6.5 mm long . . . . . . . . . . . . . . . . . . . . . . . . . . . . . . . . . . . . . . . . . . . . . . . . . . 13

12  Lowest flower at node 9 or higher, leaves usually without glandular hairs, lower floral leaves smaller than upper cauline . . . . . . . . . . . . . . . . . . . . . . . . . . . . . . . . . . . . . . . . . . . . . **nemorosa**
    Lowest flower at node 8 or lower, leaves usually with short or long glandular hairs, lower floral leaves larger than upper cauline . . . . . . . . . . . . . . . . . . . . . . . . . . . . . . . . . **arctica** subsp. **borealis**

13  Leaves densely covered with short hairs . . . . . . . . . . . . . . . . . . . . . . . . . . . . . . . . . . . . . . . . . . 14
    Leaves sparsely hairy to subglabrous (with scattered small bristles) . . . . . . . . . . . . . . . . . . . . . . . 15

14  Lowest flower at node 8 or lower, stem not more than 15 cm, lower floral leaves about as long as wide . . . . . . . . . . . . . . . . . . . . . . . . . . . . . . . . . . . . . . . . . . . . . . . . . . . . . . . . . . . . . . 36
    Lowest flower at node 9 or higher, stem to 40 cm, lower floral leaves often longer than wide . . . . . .
    . . . . . . . . . . . . . . . . . . . . . . . . . . . . . . . . . . . . . . . . . . . . . . . . . . . . . . . . . . . . . . . . . . . . . **nemorosa**

15  Stem and branches very slender, blackish, leaves strongly purple tinged, not darker beneath than above, corolla usually lilac to purple . . . . . . . . . . . . . . . . . . . . . . . . . . . . . . . . . . . . . . **micrantha**
    Stem and branches either stout or not dark in colour, leaves slightly or moderately purplish, corolla usually white . . . . . . . . . . . . . . . . . . . . . . . . . . . . . . . . . . . . . . . . . . . . . . . . . . . . . . . . 16

16  Lowest flower at node 8 or higher, leaves not darker below than above, capsule usually shorter than calyx . . . . . . . . . . . . . . . . . . . . . . . . . . . . . . . . . . . . . . . . . . . . . . . . . . . . . . . . . . **nemorosa**
    Lowest flower at node 7 or lower, leaves usually light green above and purplish beneath, capsule usually longer than the calyx . . . . . . . . . . . . . . . . . . . . . . . . . . . . . . . . . . . . . . . . . . . . . **scottica**

17  Basal pairs of teeth of lower floral leaves pointing forwards . . . . . . . . . . . . . . . . . . . . . . . . . . . . 18
    Basal pairs of teeth of lower floral leaves patent . . . . . . . . . . . . . . . . . . . . . . . . . . . . . . . . . . . . . 19

18  Teeth of lower floral leaves not much longer than wide . . . . . . . . . . . . . . . . . . . . . . . . . . . . . . . . 36
    Teeth of lower floral leaves much longer than wide . . . . . . . . . . . . . . . . . . . . . . . . . . . . . . . **confusa**

19  Lowest flower at node 10 or higher . . . . . . . . . . . . . . . . . . . . . . . . . . . . . . . . . . . . . . . . . . . . . . . 20
    Lowest flower at node 9 or lower . . . . . . . . . . . . . . . . . . . . . . . . . . . . . . . . . . . . . . . . . . . . . . . . . 21

20  Stem and branches usually slender and flexuous, leaves near base of branches usually very small . . . . . . . . . . . . . . . . . . . . . . . . . . . . . . . . . . . . . . . . . . . . . . . . . . . . . . . . . . . . . . . . **confusa**
    Stem and branches usually stout and not flexuous, leaves near base of branches not noticeably smaller than the others . . . . . . . . . . . . . . . . . . . . . . . . . . . . . . . . . . . . . . . . . . . . . . . . . . **nemorosa**

21  Leaves with numerous eglandular hairs . . . . . . . . . . . . . . . . . . . . . . . . . . . . . . . . . . . . . . . . . . . . 36
    Leaves subglabrous . . . . . . . . . . . . . . . . . . . . . . . . . . . . . . . . . . . . . . . . . . . . . . . . . . . . . . . . . . . 22

22  Capsule 5.5–7 mm, often slightly curved, inflorescence not usually quadrate . . . . . . . . **heslop-harrisonii**
    Capsule usually not more than 5.5 mm, straight, inflorescence often dense with flowers in 4 ranks (quadrate) . . . . . . . . . . . . . . . . . . . . . . . . . . . . . . . . . . . . . . . . . . . . . . . . . . . . . . . . . **tetraquetra**

23  Cauline internodes mostly at least 2.5 times leaves . . . . . . . . . . . . . . . . . . . . . . . . . . . . . . . . . . . 24
    Cauline internodes mostly less than 2.5 times leaves . . . . . . . . . . . . . . . . . . . . . . . . . . . . . . . . . . 28

24  Capsule broadly elliptical to obovate-elliptical . . . . . . . . . . . . . . . . . . . . . . . . . . . . . . . . . . . . . . 25
    Capsule oblong to narrowly elliptical . . . . . . . . . . . . . . . . . . . . . . . . . . . . . . . . . . . . . . . . . . . . . . 26

25  Teeth of lower floral leaves mostly subacute and not longer than wide, corolla 4.5–7 mm, lowest flower at node 2–5 . . . . . . . . . . . . . . . . . . . . . . . . . . . . . . . . . . . . . . . . . . . . . . . . . . . . . . **frigida**
    Teeth of lower floral leaves usually acute or acuminate and longer than broad, corolla 6.5 mm or more, lowest flower at node 4 or higher . . . . . . . . . . . . . . . . . . . . . **arctica** subsp. **borealis**

26  Upper cauline leaves elliptic-ovate to narrowly obovate . . . . . . . . . . . . . . . . . . . . . . . . . . . . **scottica**
    Upper cauline leaves suborbicular to broadly ovate or broadly obovate . . . . . . . . . . . . . . . . . . . . . 27

27  Lowest flower at node 4 or lower, lower floral leaves often much larger than the upper cauline . **frigida**
    Lowest flower at node 4 or higher, lower floral leaves scarcely larger than the upper cauline . . . . . . . . 36

28  Corolla at least 6 mm long . . . . . . . . . . . . . . . . . . . . . . . . . . . . . . . . . . . . . . . . . . . . . . . . . . . . . . . 29
    Corolla not more than 6 mm . . . . . . . . . . . . . . . . . . . . . . . . . . . . . . . . . . . . . . . . . . . . . . . . . . . . . 31

29  Teeth of lower floral leaves usually very acute, all pointing forwards . . . . . . . . . . . . . . . . . . . **confusa**

Teeth of lower floral leaves subacute to acute, the basal pairs patent............................30

30  Capsule at least as long as calyx, usually emarginate, inflorescence not quadrate ............ *frigida*
    Capsule shorter than calyx, truncate or slightly emarginate, inflorescence often quadrate ... *tetraquetra*

31  Lower floral leaves ovate to rhombic, with acute to aristate teeth, the basal pairs forward
    pointing......................................................................................... *confusa*
    Lower floral leaves broadly ovate or deltoid to suborbicular, with obtuse to subacute teeth, the
    basal pairs patent ...............................................................................32

32  Leaves with numerous eglandular hairs .....................................................33
    Leaves with few eglandular hairs ...........................................................34

33  Lowest flower at node 4 or lower, lower floral leaves often much larger than the upper cauline . *frigida*
    Lowest flower at node 4 or higher, lower floral leaves scarcely larger than the upper cauline ........36

34  Distal teeth of lower floral leaves not curved, capsule usually shorter than calyx .......... *tetraquetra*
    Distal teeth of lower floral leaves incurved, capsule at least as long as the calyx ...................35

35  Capsule usually 4.5–5.5 mm, about twice as long as wide, straight, cauline leaves only obscurely
    petiolate, margins of teeth not sinuous ......................................................... *foulaensis*
    Capsule usually 5.5–7 mm, 2–3 times as long as wide, often slightly curved, upper cauline leaves
    distinctly petiolate, margins of teeth sinuous .................................. *heslop-harrisonii*

36  Cauline leaves obovate to narrow ovate or elliptic, leaves mainly hairy towards the tip ... *campbelliae*
    Cauline leaves ovate, ovate-oblong to orbicular, leaves uniformly hairy ........................37

37  Teeth of lower floral leaves mostly broader than long, branches up to 3 pairs ............ *rotundifolia*
    Teeth of lower floral leaves mostly as broad as long, branches up to 5 pairs.......................38

38  Corolla 5.5–7 mm long, capsule usually more than twice as long as broad ................ *marshallii*
    Corolla 4.5–6 mm long, capsule not more than twice as long as broad ................... *ostenfeldii*

**E. arctica** subsp. **borealis**
(incl. **E. brevipila** auct. angl., non Burnat &
Gremli)
Common and widespread.  1–4, 6, 7, 9, 12,
13, 15, 16, 18, 19, 21, 24, 28

E. arctica subsp. borealis × confusa.  3, 9,
10, 12
E. arctica subsp. borealis × micrantha.  2, 3,
7, 12, 16
E. arctica subsp. borealis × nemorosa.  1, 3,
10, 12, 16, 21

**E. tetraquetra** (Bréb.) Arrond.
(*E. occidentalis* Wettst.)
Grassland near the sea.
Doubtful.
This species is easily confused with maritime
forms of **E.' nemorosa**, and there is no herb-
arium material which has been determined
by *Yeo* with certainty. There are records from
Lewis and St.Kilda named by *Pugsley*.  (3,
8, 9, 12, 16, 21, 28)

**E. nemorosa** (Pers.) Wallr.
Common and widespread.  1–3, 5, 8–12,
14–22, 24, 27–29

E. nemorosa × scottica.  2

**E. confusa** Pugsley
Common and widespread.  1–3, 8, 9, 12,
14, 16, 18, 19, 21

E. confusa × foulaensis.  3
E. confusa × micrantha.  2, 3, 9, 12
E. confusa × nemorosa.  1, 12, 18
E. confusa × scottica.  3, 12

**E. frigida** Pugsley
Mountain ledges, moorland.
On the higher hills generally, also St.Kilda.
South Uist, Beinn Mhor and Hecla, *M. S.
Campbell*, 1947 (BM); North Uist, North Lee,
*Hunt*, 1981 (K); North Harris, Clisham,
*Wilmott*, 1939 (CGE); Lewis, Uig, *Crabbe*,
1939 (BM) and Gress, *Wilmott*, 1939 (BM).
1, 3, 9, 16, 19, 28

E. frigida × micrantha
South Uist, Beinn Mhor, *Pankhurst*, 1983
(BM).  19

E. frigida × scottica
South Uist, Beinn Mhor, *Davis*, 1951 (LIV);
South Harris, Uamasclett, *Perring*, 1959
(CGE); Lewis, above Loch a' Chama, *Warburg*,
1946 (BM). 3, 12, 19

**E. foulaensis** Wettst.
Machair, maritime heath, mountain ledges.
Widespread. 1–3, 7–9, 12, 16, 21, 24, 28.

**E. foulaensis** × **marshallii.**
Two records determined with doubt by *Yeo*.
North Harris, Rubha Ruadh, Hushinish,
*Wilmott and Campbell*, 1939 (BM); Lewis,
Brenish, Uig, *Warburg* 1946 (BM). 3?, 9?

**E. foulaensis** × **micrantha.** 2?, 3
**E. foulaensis** × **nemorosa.** 1–3

**E. ostenfeldii** (Pugsley) P. F. Yeo
(*E. curta* auct. angl., non (Fries) Wettst.)
Doubtful. The most recent record for this
species is for 1942 and none is confirmed by
a herbarium specimen seen by *Yeo*.  (4, 9,
10, 12, 16, 28, 29)

**E. marshallii** Pugsley
Machair, maritime heath.
Local in Lewis and North Harris, with
records determined by *Pugsley* (Pugsley,
1940) from Muldoanich and Flodday. North
Harris, Rubha Ruadh, Hushinish, *M. S.
Campbell*, 1939 (BM); Lewis, south of Brenish,
*Warburg*, 1946 (BM), head of Eoropie Bay,
*Wilmott* 1946 (BM).   1, 3, 9, 22, 23

**E. marshallii** × **nemorosa.** Butt of Lewis at
Eoropie, *M. S. Campbell*, 1946 (BM) and
Lionel, *M. S. Campbell*, 1937 (BM).   1

**E. rotundifolia** Pugsley
Machair, rare, only recorded from Branahuie
near Stornoway, *M. S. Campbell, Warburg &
Crabbe*, 1946 (BM).   1
This is a protected species, listed in the Red
Data Book, Perring and Farrell (1983).

**E. campbelliae** Pugsley
Damp coastal moorland, endemic to Lewis.
1–4
This is also a protected species, listed in the
Red Data Book, Perring and Farrell (1983).

**E. campbelliae** × **confusa.**  3
**E. campbelliae** × **marshalii.**  3
**E. campbelliae** × **micrantha.**  2
**E. campbelliae** × **nemorosa.**  2

**E. micrantha** Reichb.
Damp moorland and heath.
Common from Mingulay to Lewis and St
Kilda.   1–5, 7, 9–12, 16–19, 21, 22, 24, 28

**E. micrantha** × **nemorosa.**  2, 3, 9, 12, 15,
21.
**E. micrantha** × **scottica.** According to
*Silverside*, commoner on the moorlands of
Lewis than its parents, and under-recorded.
2, 9, 21

**E. scottica** Wettst.
Damp moorland and heath.
Common from South Uist to Lewis and St
Kilda.   1–3, 5, 7–9, 12, 15–19, 21, 22, 24, 28

**E. heslop-harrisonii** Pugsley
Upper margins of saltmarshes.
Rare. South Harris, Luskentyre, one small
population discovered by *Silverside* in 1982.
12
This is also a protected species, listed in the
Red Data Book, Perring and Farrell (1983).

**Odontites** Ludw.
**O. verna** (Bell.) Dumort.
Red Bartsia.
Modhalan Coitcheann.
Native.
Waste ground, roadsides and damp places.
From Berneray and Mingulay to Lewis and
the Shiant Is.
The taxonomy of this genus is somewhat
confused; we have followed the work of
Snogerup (1982) (see Rich and Rich, 1988).
1–4, 6, 7, 9, 10, 12, 14–25, 27, 29

**O. vulgaris** Moench
(subsp. **serotina** (Dumort.) Corb. (= *O. serotina* Dumort.) and subsp. **pumila** (Nordst.)
Pedersen).   3, 12, 15, 16, 18, 19

**O. litoralis** Fries.
(**O. verna** subsp. **litoralis** (Fries) Nyman).
1–3, 9, 10, 16, 17, 19, 21

*OROBANCHACEAE*
**Orobanche** L.
**O. alba** Steph. ex Willd.
Thyme Broomrape.
Siorralach.
Native.
This parasite grows on **Thymus**, and in the
Outer Hebrides at least, is bright red, in
spite of its name.
From Fuday and Eriskay to South Harris.
12, 18–21

*LENTIBULARIACEAE*
**Pinguicula** L.
**P. lusitanica** L.

Pale Butterwort.
Mòthan Beag Bàn.
Native.
Bogs and flushes.
Rare to frequent from Mingulay to Lewis.
1–7, 9, 12, 13, 16, 18–24, 27

**P. vulgaris** L.
Common Butterwort.
Mòthan.
Native.
Damp moorland.
From Berneray and Mingulay to Lewis, Shiant Is. and St Kilda.   1–7, 9–24, 27, 28

**Utricularia** L.
**U. australis** R. Br.
*U. neglecta* Lehm.
Bladderwort.
Lus nam Balgan Mòr.*
Native.
Uncommon. There are 2 specimens in BM; South Uist, Daliburgh *M. S. Campbell*, *Warburg* and *Crabbe* and Lewis, Uig, Loch a' Chama *Crabbe*, there are also 2 literature records, North Uist, Lochmaddy *Shoolbred* (1895); *Ribbons* recorded it on the BSBI excursion to Stornoway (Copping, 1976). Field records from Barra, the Uists, South Harris and the Eye peninsula.   3, 6, 7, 12, 16, 18, 19, 21

**U. intermedia** Hayne
Intermediate Bladderwort.
Lus nam Balgan Meadhanach.*
Native.
Mesotrophic lochs.
According to Thor (1988) this should be split into 3 species, **U. intermedia** s. s., **U. stygia** Thor and **U. ochroleuca** R. Hartm. For the differences, see Rich and Rich (1988). There is one record for **O. ochroleuca** from Benbecula, *Horsman* 1988 det. *P. Taylor*. Accurate identification appears to need copious fresh material, and it has not been possible to separate these species in the herbarium.
North Uist to West Lewis.   1, 3, 7, 9, 11, 12, 16, 18–22

**U. minor** L.
Lesser Bladderwort.
Lus nam Balgan Beag.*
Native.
The commonest species, from Sandray to Lewis.   2–5, 7–9, 12, 15, 16, 18, 19, 21, 23

*LABIATAE*
[LAMIACEAE]
**Mentha** L.
**M. arvensis** L.
Corn Mint.
Meannt an Arbhair.
Native.
Very occasional, the only specimen seen is; coast, Branahuie, Stornoway, Lewis *M. S. Campbell, Warburg* and *Crabbe* (BM) but there are various literature reports from Harrison (1941 and 1956), Harrison et al. (1942a), on Mingulay, North Uist and Harris.   1, ?12, 16, ?24

**M. aquatica** L.
Water Mint.
Meannt an Uisge.
Native.
Mingulay to Lewis and the Shiant Is. though apparently rare in the North.   2, 6, 12, 15, 16, 18, 19, 21–24, 27

**M. × piperata** L.
*M. aquatica × spicata*
Peppermint.
Introduction.
South Harris and Lewis; Luskentyre, South Harris *Wilmott* and *M. S. Campbell* (BM) 1939; Meavaig, Uig, Lewis *Wilmott* (BM) 1946; Ensay, *J. K. Morton*, 1958.   3, 6, 12, 14

**M. spicata** L.
Spear Mint.
Meannt Gàrraidh.
Denizen.
Nunton, Benbecula *Crabbe*; Dalebeg, Barvas, Lewis *Wilmott* and *M. S. Campbell*; ?Meavaig, Uig, Lewis (all BM), Harrison (1941a) also recorded it from South Uist and Barra.   2, 3, 18, 19, 21

**M. × villosa** Hudson
*M. niliaca* auct. *M. spicata × suaveolens*
Apple-mint.
Probably introduced.
Lewis, Kirivick, Harrison (1957); North Uist, Boreray, *J. W. Campbell* (BM) 1937 determined by *Still*.   2, 15

**Lycopus** L.
**L. europaeus** L.
Gipsywort.
Feòran Curraidh.
Native.
Marshy places.
Mingulay to Harris.   9, 10, 12, 19–21, 24

**Origanum** L.
**O. vulgare** L.
Marjoram.
Oragan.
Introduced.
Lewis, Stornoway Castle woods, Harrison (1948) and Cunningham (1978).   5

**Thymus** L.
**T. pulegioides** L.
Large Thyme.
There are 4 specimens identified as this in BM; 3 are a large form of the next species and the other, Valtos Glen, Uig, Lewis *Wilmott* 1939 is sterile but might possibly be this species.

**T. praecox** Opiz
subsp. **arcticus** (Durand) Jalas
*T. drucei* Ronniger
Wild Thyme.
Lus na Macraidh, Lus an Rìgh.
Native.
Dry moorland, banks and rocky places.
Mingulay to Lewis, St Kilda to Shiant Is. 1–7, 9–24, 27–29

**Prunella** L.
**P. vulgaris** L.
Selfheal.
Dubhan Ceann-chòsach.
Native.
Machair, damp grassland, lochsides and roadsides.
Mingulay and Berneray to Butt of Lewis and St Kilda to Shiant Is.   1–24, 27, 28, 29

**Stachys** L.
**S. arvensis** (L.) L.
Field Woundwort.
Creuchd-lus Arbhair.**
Doubtfully native.
First recorded by *Findlayson* from Mingulay fide Bennett (1890); weed in oats above Castlebay, Barra *Wilmott* 1938 (BM); North Uist, rye field at Balmartin *J. W. Clark* 1981; there are literature records from Mingulay, Barra and South Uist, King's College (1939) and Harrison (1941a).   16, 19, 21, 24

**S. palustris** L.
Marsh Woundwort.
Brisgean nan Caorach.
Native.
Wet ditches.
Barra to Lewis.   1–3, 6, 7, 9, 10, 12, 15–17, 19, 21

**S. sylvatica** L.
Hedge Woundwort.
Lus nan Sgor.
Native.
Rare, first recorded by Balfour and Babington from Rodel, South Harris then collected there by *M. S. Campbell* (BM) and *J. W. H. Harrison* (K), also from Kyles Stuley, South Uist *J. W. H. Harrison* (K); there are literature records from Lochboisdale, South Uist (Harrison, 1941a) and Barra *Conacher* (1980). 12, 19, 21

**S.** × **ambigua** Smith
*S. palustris* × *sylvatica*
Native or introduced.
Barra, North Uist, South Harris and Lewis, often growing in the absence of the parents, and so suspected of being introduced.   2, 12, 16, 21

**Lamium** L.
**L. amplexicaule** L.
Henbit Dead-nettle.
Deanntag Chearc.
Native.
Arable land.
Barra to Harris.   11, 12, 15, 16, 19, 21

**L. moluccellifolium** Fries
Northern Dead-nettle.
Deanntag Thuathach.
Possibly native.
This species can normally be distinguished from the other red flowered dead-nettles by the size of the calyx (8–12 mm, not 5–7 mm) and the teeth of the calyx being longer than the tube. However if **L. amplexicaule** has non-cleistogamous flowers these can approach this size.
Mingulay to Lewis.   1–4, 8, 9, 11, 12, 16, 18, 19, 21, 22, 24

**L. hybridum** Vill.
Cut-leaved Dead-nettle.
Deanntag Gheàrrte.
Native.
Arable land.
Vatersay to Lewis.   3, 8, 12, 15, 16, 18, 19, 21, 22

**L. purpureum** L.
Red Dead-nettle.
Deanntag Dhearg.
Native.
Arable land.
Vatersay to Lewis.   3, 5–7, 10, 12, 15–22

**L. album** L.
White Dead-nettle.
Deanntag Bhàn, Teanga Mhìn.
Probably introduced.
One record only from Balelone, North Uist
by *J. W. Clark* and *Murray*, 1982.   16.

**Galeopsis** L.
**G. tetrahit** L.
Common Hemp-nettle.
Deanntag Lìn, Gath Dubh.
Native.
Arable land.
The closely related **G. bifida** Boenn. is also
recorded, but most of the records are for the
aggregate species.
Mingulay to Lewis and St Kilda.   2, 3, 5–7,
9–12, 15–21, 24, 28

**Scutellaria** L.
**S. galericulata** L.
Skullcap.
Currac-claiginn Mòr.*
Native.
Margins of lochs, sea beaches.
South Uist to Harris.   9, 12, 16, 19
A distinctive densely-branched ecotype
occurs on shingle beaches.

**S. minor** Hudson
Lesser Skullcap.
Currac-claiginn Beag.*
Native.
Damp moorland and flushes.
Barra to Lewis.   1–5, 7, 9–12, 15, 16, 18, 19,
21–23

**Teucrium** L.
**T. scorodonia** L.
Wood Sage.
Sàisde Coille.
Native.
Rocky moorlands
Barra to Lewis and possibly St Kilda (as
'sage' *Martin Martin* 1703).   1, 3, 4, 7, 9, 12,
14, 16, 19, 21, ?28

**Ajuga** L.
**A. reptans** L.
Bugle.
Meacan Dubh Fiadhain.
Native.
Rare, except in South Uist. Lewis, Stornoway
Castle grounds; South Uist in various locali-
ties; Berneray (Harris), Sandray and Shiant
Isles.   5, 15, 19, 23, 27

**A. pyramidalis** L.
Pyramidal Bugle.
Native.
Rock crevices and ravines.
Rare. Barra, South Uist, South Harris and
Lewis near Stornoway.   1, 5, 12, 19, 21

PLANTAGINALES
*PLANTAGINACEAE*
**Plantago** L.
**P. major** L.
Greater Plantain.
Cuach Phàdraig.
Native.
Waste places, roadsides and tracks.
Berneray and Mingulay to Lewis, St Kilda to
Shiant Is. and to North Rona.   1–7, 9, 10,
12, 15–22, 24, 25, 27–29

**P. lanceolata** L.
Ribwort Plantain.
Slàn-lus.
Native.
Waste places, machair and grassland.
Berneray and Mingulay to Lewis, St Kilda to
Shiant Is.   1–24, 27, 28

**P. maritima** L.
Sea Plantain.
Slàn-lus na Mara.
Native.
Salt marshes, waste places near the sea and
machair.
Probably in all zones.   1–24, 27–29

**P. coronopus** L.
Buck's-horn Plantain.
Adhairc Fèidh.**
Native.
Machair, salt marshes, roadsides and rocky
places.
Probably in all zones.   1–25, 27–29

**Littorella** Bergius
**L. uniflora** (L.) Ascherson
Shoreweed.
Lus Bòrd an Locha.*
Native.
Oligotrophic loch margins.
Berneray and Mingulay to Lewis.   2–7, 9,
12, 13, 15–24, 29

CAMPANULALES
*CAMPANULACEAE*
**Campanula** L.
**C. rotundifolia** L.
Bluebell (Harebell in England).

Currac-cuthaige, Am Flùran/Plùran Cluige-
anach.
Native.
Machair and grassland.
Pabbay to Lewis.   3–5, 11, 12, 15, 16, 18–24,
29

LOBELIACEAE
Lobelia L.
L. dortmanna L.
Water Lobelia.
Flùr an Lochain.
Native.
Oligotrophic lochs.
Barra to Lewis.   1–5, 7, 9–12, 15, 16–19, 21

RUBIALES
RUBIACEAE
Sherardia L.
S. arvensis L.
Field Madder.
Màdar na Machrach.
Native.
Arable land and machair.
The Uists and Benbecula with a single collec-
tion from Coll, Lewis.   1, 16, 18, 19

Cruciata laevipes Opiz
(Galium cruciata (L.) Scop.)
Crosswort.
Doubtfully native.
Rare, only one record. South Uist, field
record from square NF72 Nevison 1956. Re-
quires confirmation.   19

Galium L.
G. verum L.
Lady's Bedstraw.
Lus an Leasaich, Rùin (Ruadhain).
Native.
Machair, roadsides and dry places.
Mingulay to Lewis.   1–4, 6, 7, 12–16, 18–24,
29

G. saxatile L.
Heath Bedstraw.
Màdar Fraoich.
Native.
Moorland and rocky places.
Mingulay to Lewis and St Kilda to Shiant
Is.   3–7, 9–12, 15–21, 24, 27, 28

G. sterneri Ehrend.
(G. pumilum auct., pro parte)
Limestone Bedstraw.
Native.
Recorded in literature twice; narrow sea-loch

near Lochmaddy, North Uist Shoolbred (1895);
Glen Skeaudale, North Harris (Harrison,
1956).
No specimens seen and the mainland distribution makes this identification unlikely; the
records probably only refer to a form of the
preceding species.

G. palustre L.
Common Marsh-bedstraw.
Màdar Lèana.
Native.
Marshes and flushes.
Barra to Lewis and Shiant Is.   1, 2, 5, 6, 11,
12, 14–22, 27, 29

G. uliginosum L.
Fen Bedstraw.
As with G. sterneri this species has been
only recorded in literature; South Uist
(Bennett, 1889); Kildonan, South Uist and
Vatersay (Harrison, 1941a). There is also a
field record for Rodel, South Harris. There
are no specimens and the records need to be
confirmed.   12, 19, 22

G. aparine L.
Cleavers.
Garbh-lus.
Native.
Rocks, shingle and machair.
Mingulay to Lewis.   2–7, 9, 12–16, 18–24,
27, 29

CAPRIFOLIACEAE
Sambucus L.
S. ebulus L.
Dwarf Elder.
Planted.
Field record for South Uist, Grogarry House,
1983.   19

S. nigra L.
Elder.
Droman.
Native and planted.
Garden hedges and scrub.
Barra to Harris and St Kilda.   5–7, 9, 10, 12,
14, 15–19, 21, 28

Symphoricarpus Duh.
S. albus (L.) S. F. Blake
var. laevigatus (Fernald) S. F. Blake
(S. rivularis Suksdorf)
Snowberry.
Planted.
Lewis, Stornoway Castle, Cunningham

(1978); Eye peninsula, field record *Roe* 1959.
5, 6

**Lonicera** L.
**L. periclymenum** L.
Honeysuckle.
Iadh-shlat, Lus na Meala.
Native.
Sea and lochside cliffs, islands and stream
ravines.
Mingulay to Lewis and St Kilda.    1–5, 7,
9–13, 16–24, 27, 28
A variable plant, especially in the shape of
the leaf and its hairiness.*J. W. H. Harrison*
recorded var. *quercifolia* Aiton from Rueval,
Benbecula and described var. *clarki* (Harrison,
1938a) and var. *angustifolia* from the islands
but these represent minor variants only, not
worthy of separation.

*VALERIANACEAE*
**Valerianella** Miller
**V. locusta** (L.) Laterrade
Common Cornsalad.
Leiteis an Uain.
Native
Arable fields and dunes.
var. **dunensis** D. E. Allen (subsp. *dunensis*
(D. E. Allen) P. D. Sell) is recorded from
Machair Robach, North Uist *Murray* 1974 (E)
Barra to Lewis.    1, 12, 16, 18–21

**V. dentata** (L.) Pollich
Narrow-fruited Cornsalad.
Native.
Recorded by *Shoolbred* (1895) from near
Tigharry, North Uist; specimen in BM from
Vallay Island, North Uist *Shoolbred* 1898 (BM)
conf. *Mullin*.    16

**Valeriana** L.
**V. officinalis** L.
Common Valerian.
Carthan Curaidh.
Native.
Barra to Lewis.    3, 9, 12, 19, 21

*DIPSACACEAE*
**Knautia** L.
**K. arvensis** (L.) Coulter
Field Scabious.
Gille Guirmein.
Possibly native.
Machair.
Known from three literature records 'Outer
Islands, H. H. Harrison (1939); Vatersay
(Harrison, 1941a), and from Little Bernera

(Crummy, 1982). Collected on machair at
Udal in North Uist by *Murray* in 1969.    4,
16, 22

**Succisa** Haller
**S. pratensis** Moench
Devil's-bit Scabious.
Greim an Diabhail, Ura-bhallach.
Native.
Moorland, mountains, machair and pastures.
Probably in all zones.    1–24, 27–29

ASTERALES
*COMPOSITAE*
*[ASTERACEAE]*
**Senecio** L.
**S. jacobaea** L.
Common Ragwort.
Buaghallan.
Native.
Pastures, banks and machair.
Coastal populations are particularly prone to
variation. In Rich and Rich (1988) subsp.
**dunensis** (Dumort.) Kadereit & P. D. Sell is
distinguished by the more or less complete
absence of ray florets. There is one specimen
from Scarp, *W. S. Duncan*, 1891 (BM). Var.
**condensatus** Druce does have ray florets but
its stems are markedly swollen below the
basal leaves; there is one record from the
beach at Teanamachar, Baleshare, North
Uist by *Preston* 1987 det. *Sell*. A conspicu-
ously glandular variant has been collected
on machair at Daleburgh, South Uist.
Probably in all zones.    1–24, 27–29

**S. aquaticus** Hill
Marsh Ragwort.
Caoibhreachan.
Native.
Marshes, wet pastures and ditches.
Berneray and Mingulay to Lewis, St Kilda to
Shiant Is.    1, 4, 6, 11–13, 15, 16, 18–24, 27,
28

**S. × ostenfeldii** Druce
*S. aquaticus × jacobaea*
Recorded by Harrison (1941a) from
Benbecula, South Uist and Vatersay, with no
specimens; field record from Loch Fasgro,
Lewis *Noltie* 1987; there is a specimen from
Howmore, South Uist *Pankhurst* 1983 (BM).
2, 18, 19, 22

**S. vulgaris** L.
Groundsel.
Grunnasg.

Native or colonist.
Arable land and dunes.
Mingulay to Lewis and St Kilda.   3–7, 10, 12, 15–22, 24, 27–29

**Tussilago** L.
**T. farfara** L.
Colt's-foot.
Cluas Liath.
Native.
Waste ground, streamsides and unstable slopes near the sea.
Mingulay to Lewis (Stornoway Castle and Eye Peninsula).   5–7, 12, 13, 15, 16, 20, 21, 23, 24

**Petasites** Miller
**P. hybridus** (L.) P. Gaertner, B.Meyer & Scherb.
Butterbur.
Gallan Mòr.
Native.
Damp pastures, damp slopes by the sea and machair. Currie (1977) and again Naylor and Cumming (1982) describe invasive behaviour of this species in machair.
Mingulay, Barra, North Uist and Lewis. 1–5, 13–16, 20–22, 24

**P. albus** (L.) Gaertner
White Butterbur.
Gallan Mòr Bàn.
Introduction.
Castle Park, Stornoway, Lewis *M. S. Campbell* 1938 (BM).   5

**P. fragrans** (Vill.) C. Presl
Winter Heliotrope.
Gallan Mòr Cùbhraidh.
Introduction.
Recorded from Castle Park, Stornoway, Lewis *Conacher* and *Ribbons* (Copping, 1977) and Vatersay *Conacher* (1980).   5, 22

Homogyne Cass.
H. alpina (L.) Cass.
Purple Colt's-foot.
A record from Beinn Mhor, South Uist was published by *Spence* in Proc.BSBI 2:374 (1957). This is now known to have been planted by students as a hoax in about 1955, and does not appear to have survived (C. D. Preston, pers.comm.).

**Inula** L.
**I. helenium** L.
Elecampane.

Ailleann.
Introduction.
Gardens, waste ground and pastures near houses.
Barra to Lewis.   3, 4, 11, 14, 16, 18, 19, 21

**Gnaphalium** L.
**G. sylvaticum** L.
Heath Cudweed.
Cnàmh-lus Mòintich.
Native.
Field record from near L. Odhair, Parc, *Chater* and *Pankhurst*, 1980; Benbecula *Shoolbred* 1894 (BM); near Lochboisdale, South Uist (Harrison, 1941a).   7, 18, 19

**G. uliginosum** L.
Marsh Cudweed.
Cnàmh-lus Lèana.
Native.
Waste ground and tracks.
Barra to Lewis.   2, 5–7, 15–21

**Anaphalis** DC.
**A. margaritacea** (L.) Bentham
Pearly Everlasting.
Introduction.
North Tolsta, Lewis *Pankhurst* and *Chater* (BM).   1

**Antennaria** Gaertner
**A. dioica** (L.) Gaertner
Mountain Everlasting.
Spòg Cait.
Native. Rocky knolls on moorland.
Berneray and Mingulay to Lewis and St Kilda.   1–4, 7, 9, 10, 12, 13, 15, 16, 18–22, 24, 28

**Solidago** L.
**S. virgaurea** L.
Goldenrod.
Slat Oir.
Native.
Rocky loch and streamsides.
Mingulay to Lewis and the Shiant Is.   1, 3–5, 7–10, 12, 13, 16–19, 21–24, 27

**Aster** L.
**A. tripolium** L.
Sea Aster.
Neòinean Sàilein.
Native.
Saltmarshes.
South Uist to Lewis, but surprisingly, absent from Harris.   1, 7, 14–16, 18, 19
var. **arctium** Fries has been recorded from

North Uist and from Berneray but appears to be only a dwarf form.

**Bellis** L.
**B. perennis** L.
Daisy.
Neòinean.
Native.
Machair, roadsides, pasture and waste ground.
Probably in all zones.   1–7, 9–25, 27–29

**Anthemis** L.
**A. cotula** L.
Stinking Chamomile.
Sineal.
Probably an introduction.
Waste ground.
Benbecula and the Uists.   16, 18, 19

**Chamaemelum** Miller
**C. nobile** (L.) All.
Chamomile.
Cama-bhil.
Probably introduced.
Only recorded from Luskentyre Banks, South Harris *J. W. H. Harrison* et al., 1941a

**Achillea** L.
**A. ptarmica** L.
Sneezewort.
Cruaidh-lus.
Native.
Marshes and flushes.
Barra to Lewis and Shiant Is.   1, 3, 6, 7, 10–12, 15, 16, 18–22, 27

**A. millefolium** L.
Yarrow.
Eàrr-thalmhainn.
Native.
Machair, roadsides, dry grasslands and banks.
Berneray and Mingulay to Lewis and North Rona, St Kilda to Shiant Is.   1–7, 9–24, 25, 27–29

The taxonomy of the next genus is difficult and possibly the species would be best treated as subspecies. Records have certainly been confused.

**Tripleurospermum** Schultz Bip.
**T. maritimum** (L.) Koch
*Matricaria maritima* L.
Sea Mayweed.
Buidheag an Arbhair.
Native.
Most plants have dark margins to the involucral bracts, and the more extreme have been named as var. *phaeocephala* Ruprecht, Beeby (1887), now known as subsp. **phaeocephalum**. Nevertheless, Hebridean plants do not appear to show the other characters of this arctic subspecies.
Berneray and Mingulay to Lewis and North Rona, St Kilda to Shiant Is.   1–13, 15–25, 27–29

**T. inodorum** (L.) Schultz Bip.
*Maticaria perforata* Mérat
Scentless Mayweed.
Native.
Apparently only recorded from the Sula Sgeir and possibly North Rona *Stewart* 1932 and *Atkinson* (1940).   25

**Matricaria** L.
**M. recutita** L.
*Chamomilla recutita* (L.) Rauschert
Scented Mayweed.
Native.
Rare. Benbecula, field record in square NF85, *U. K. Duncan* 1960. Requires confirmation.   18

**M. matricarioides** (Less.) Porter
*Chamomilla suaveolens* (Pursh) Buchenau
Pineappleweed.
Lus Anainn.**
Introduction.
Waste ground, arable fields and tracks.
Mingulay to Lewis. Harrison (1941a) states that it is found in all inhabited islands except Eriskay.   1–3, 5–7, 10, 12, 13, 15–22, 24, 29

**Chrysanthemum** L.
**C. segetum** L.
Corn Marigold.
Bile Bhuidhe.
Native.
Arable fields and machair.
Barra to Lewis, Monach Is. and St Kilda.   3, 4, 6, 7, 12, 15–19, 21, 22, 28, 29

**Leucanthemum** Miller
**L. vulgare** Lam.
*Chrysanthemum leucanthemum* L.
Oxeye Daisy.
Neòinean Mòr.
Native.
Roadsides and machair.
Barra to Harris.   9, 12, 16, 18, 19, 21, 22

**Tanacetum** L.
**T. parthenium** (L.) Schultz Bip.
*Chrysanthemum parthenium* (L.) Bernh.
Feverfew.
Meadh Duach.
Introduction.
Near houses and waste ground.
Barra, Castlebay, *Pankhurst* 1983; Lewis,
field record in square NB33 c.1950.    2/5, 21

**T. vulgare** L.
*Chrsanthemum vulgare* (L.) Bernh.
Tansy.
Lus na F01ange.
Probably introduced.
Roadsides near houses.
Barra to Lewis.    2–4, 6, 7, 9, 12, 15, 16, 18,
19, 21

**Artemisia** L.
**A. vulgaris** L.
Mugwort.
Liath-lus.
Native.
Waste places.
Barra to Lewis and St Kilda.    2–4, 7, 9, 11,
12, 15–22, 24, 28, 29

**A. absinthium** L.
Wormwood.
Burmaid.
Introduction.
In or near gardens.
South Uist, field record 1951; Great Bernera,
Breaclete, Harrison (1957) and Conacher
(1980).    4, 19

Carlina vulgaris L.
Carline Thistle
Reported from machair at Grogary by
Dickinson (1968). Doubtful, and needs to be
confirmed.

**Arctium** L.
**A. minus** Bernh.
Lesser Burdock.
Leadan Liosda, Cliadan.
Native.
Waste places, pastures and roadsides.
Mingulay to Lewis.    2–4, 6, 7, 9, 11–24, 29
It has been suggested that some records
from South Uist, Eriskay and Little Bernera
to Lewis to Pabbay by *J. W. H. Harrison*
represent **A. nemorosum** Lej. but no speci-
mens have been seen. The BSBI Atlas
Supplement shows all the records from the
Outer Hebrides as belonging to this latter

species or subspecies (*A. minus* L. subsp.
*nemorosum* (Lej.) Syme).

**Cirsium** Miller
**C. vulgare** (Savi) Ten.
Spear Thistle.
Cluaran Deilgneach.
Native.
Waste places, pastures and roadsides.
Mingulay to Lewis and St Kilda.    2–7, 9,
11–24, 27–29

**C. palustre** (L.) Scop.
Marsh Thistle.
Cluaran Lèana.
Native.
Moorland and flushes.
Barra to Lewis.    1, 5, 7, 11, 12, 16–19, 21

**C. arvense** (L.) Scop.
Creeping Thistle.
Fòthannan Achaidh.
Native.
Roadsides, waste places and poor pastures.
None of the many described varieties of this
species has been recorded in the Outer
Hebrides.
Berneray and Mingulay to Lewis and St.
Kilda.    2–6, 10–24, 28, 29

**C. helenioides** (L.) Hill
*C. heterophyllum* (L.) Hill
Melancholy Thistle.
Cluas an Fhèidh.
Native.
Rare. Recorded from Lochmaddy, North
Uist *J. W. Campbell* 1947 and a white form
from south of Loch Corodale, South Uist,
*Braithwaite*, 1983. There is a specimen from
between Scarista and Borve, South Harris
*M. S. Duncan* 1891 (BM), and Harrison (1957)
probably refers to the same locality.    12, 16,
19

**Saussurea** DC.
**S. alpina** (L.) DC.
Alpine Saw-wort.
Native.
Summit rocks and gullies.
Only in the the areas of continous high
ground, Harris and Lewis and South Uist. 3,
9, 12, 19

**Centaurea** L.
**C. scabiosa** L.
Greater Knapweed.
Cnapan Dubh Mòr.

Possibly introduced.
There is only a single specimen of this plant from Newton, North Uist *J. W. Campbell* (Campbell, 1937b) (BM). The specimen is correct; it could be an escape from a cottage garden.

**C. nigra** L.
Common Knapweed.
Cnapan Dubh.
Native.
Pastures, machair, cliffs and waste places.
In the British Isles this is commonly split into 2 subspecies dependant on the degree of swelling below the capitula but there is a great deal of intergrading. The northern subspecies is subsp. **nigra** and all records from the Outer Hebrides refer to that. Specimens determined as subsp. **nemoralis** (Jordan) Gugler by *J. W. H. Harrison* et al. are not correct.
Mingulay to Lewis and Shiant Is.   3–5, 7, 9–13, 15–24, 29

**Cichorium** L.
**C. intybus** L.
Chicory.
Lus an t-Siùcair.
Introduced.
Literature records from machair, Luskentyre, South Harris *M. S. Campbell* and *Wilmott* (Campbell, 1940) and Harrison (1956); North Uist, in a field between Balmore and Cnoc Hasten for many years, *J. W. Clark*.   12, 16.

**Lapsana** L.
**L. communis** L.
Nipplewort.
Duilleag-bhràghad.
Native.
Scrub by streams.
Barra to Lewis.   3, 7, 9, 12, 16, 21

**Hypochoeris** L.
**H. radicata** L.
Cat's-ear.
Cluas Cait.*
Native.
Waste places and heathy pastures.
First record by *Shoolbred* in 1894.
Vatersay to Lewis.   1–5, 7, 9–12, 15, 16, 18–22, 27, 29

**Leontodon** L.
**L. autumnalis** L.
Autumn Hawkbit.
Native.

Caisearbhan Coitcheann.
Moorland, machair, pasture, flushes and waste places.
Subsp. **pratensis** (Koch) Arcangeli has been recorded from St. Kilda; Skigersta and Geshader, Lewis; and Daleburgh, South Uist. It is distinguished mainly by dense dark long hairs on the involucre, but the species is highly variable and the status of the subspecies is doubtful.
Probably in all zones.   1–25, 27–29

**L. hispidus** L.
Rough Hawkbit.
Recorded by Harrison (1941a) from South Uist, Eriskay and Fuday. There are no herbarium specimens and these records need to be confirmed.

**L. taraxacoides** (Vill.) Mérat
(*L. leysseri* Beck)
Lesser Hawkbit.
Caisearbhan as Lugha.
Native.
Berie Sands, Lewis *Crabbe* 1939 (BM); Harrison et al. (1939) and Harrison (1941a) record this species from South Uist, Eriskay, and Fuday. The Royal Botanic Garden, Edinburgh (1983) record it from near various lochs in Benbecula and South Uist.   3, 18, 19, 20, 21

**Sonchus** L.
**S. arvensis** L.
Perennial Sow-thistle.
Bliochd Fochain.
Native.
Arable fields and seashore shingle.
Mingulay to Lewis.   3, 5, 12, 13, 15–24

**S. oleraceus** (L.) Hill
Smooth Sow-thistle.
Bainne Muice.
Native.
Waste places.
Mingulay to Lewis and St Kilda.   1, 2, 12, 15, 16, 18, 20, 21, 24, 27, 28

**S. asper** (L.) Hill
Prickly Sow-thistle.
Searbhan Muice.
Native.
Arable Fields and waste places.
Berneray and Mingulay to Lewis and St Kilda.   3–10, 12, 13, 15–19, 21–24, 28, 29

**Hieracium** L.
Hawkweed.

Lus na Seabhaig.
Native.
Rocky banks, cliffs and crags, ravines, walls.
Throughout the larger islands, widespread but scarce, and in our experience, usually in small populations. It can only survive when it is out of the reach of grazing animals. 1–5, 7–10, 12, 16, 18, 19, 21, 23

The genus *Hieracium* is apomictic and more than 200 microspecies are known from the British Isles. The following account is based mainly on material determined by either W. H. Pugsley or by P. D. Sell and C. West at the British Museum or Cambridge. Because of the critical nature of the genus and the changes in its taxonomic treatment over time, the older records cannot be accepted unless supported by herbarium material. The nomenclature used here follows P. D. Sell and C. West in Perring and Sell (1968). Hawkweeds are generally calcicolous montane plants, except for certain weedy species. Since the soils of the Outer Hebrides are generally acid and unsuitable for *Hieracium*, it is surprising that there are so many species recorded, and that several of them are of local distribution or even endemic (*H. scarpicum*). The section *Alpina* is completely absent, as is the predominantly northern section *Alpestria*, which is so richly represented in the Shetlands (Scott and Palmer, 1987). Weedy species, such as *H. exotericum*, are also largely absent.

It is worth listing the species recorded from some of the richer localities.

1) grounds of Stornoway Castle. *H. flocculosum, sparsifolium, strictiforme* and *subcrocatum*.

2) North and South Lee hills, North Uist. *argenteum, rivale, uisticola* and *vulgatum*.

3) Ben Eaval, also in North Uist. *anglicum, argenteum, caledonicum, ebudicum, strictiforme* and *uistense*.

4) Allt Volagir, a ravine on Beinn Mhor, South Uist, with *H. caesiomurorum, eboracense, hebridense, euprepes, orimeles, rubiginosum, shoolbredii, sparsifolium* and *uiginskyense*.

In the following key, certain terms are used which may not be familiar, or are used in different senses for other plant groups;

**acladium** refers to the terminal flower head in the inflorescence, and sometimes its peduncle as well;

**capitulum** measurements refer to the diameter of the open head;

**geminate** refers to flower heads occurring in pairs (literally twinned). This usually refers to the acladium and a neighbouring head when they both have short and more or less equal peduncles, so that two heads occur close together;

**senescent** refers to tufts of stellate (floccose) hairs at the tips of the phyllaries. This has the effect of giving the bud a white tip, which fades to brown in herbarium specimens;

**stylose** refers to flower heads with disc florets only and without ray florets.

### Key to *Hieracium* in the Outer Hebrides

1  Stem leaves 6 to 30 . . . . . . . . . . . . . . . . . . . . . . . . . . . . . . . . . . . . . . . . . . . . . . . . . . . . . . . . . . . . . . .2
   Stem leaves 0 or 1 . . . . . . . . . . . . . . . . . . . . . . . . . . . . . . . . . . . . . . . . . . . . . . . . . . . . . . . . . . . . . . .13
   Stem leaves 2 to 5 . . . . . . . . . . . . . . . . . . . . . . . . . . . . . . . . . . . . . . . . . . . . . . . . . . . . . . . . . . . . . . .31

2  Phyllaries epilose, or with few pilose hairs . . . . . . . . . . . . . . . . . . . . . . . . . . . . . . . . . . . . . . . . .3
   Phyllaries densely pilose  . . . . . . . . . . . . . . . . . . . . . . . . . . . . . . . . . . . . . . . . . . . . . . . . . . . . . . . . . .6

3  Phyllaries densely glandular . . . . . . . . . . . . . . . . . . . . . . . . . . . . . . . . . . . . . . . . . . . . . . . . . . . . . . .4
   Phyllaries with few glandular hairs  . . . . . . . . . . . . . . . . . . . . . . . . . . . . . . . . . . . . . . . . . . . . . . . . .5

4  Leaves  dark green, lingulate or oblong-lanceolate, subentire, peduncles medium or long, buds
      truncate at base, fresh styles livid . . . . . . . . . . . . . . . . . . . . . . . . . . . . . . . . . . . . . . . . . *H. strictiforme*
   Leaves yellowish or dull green, elliptic-lanceolate, denticulate to distantly dentate, peduncles
      short, bud rounded below, fresh styles pure yellow . . . . . . . . . . . . . . . . . . . . . . . . . . . . *H. latobrigorum*

5  Leaves dark green, glabrous above, margins scaberulous, stem leaves oblong, sub-entire to
      denticulate, buds truncate at base, fresh styles fuscous . . . . . . . . . . . . . . . . . . . . . . . . . . *H. maritimum*
   Leaves yellowish or dull green, glabrescent or pilose above, margins shortly ciliate, stem leaves
      elliptic or lanceolate, stem leaves dentate, buds rounded below, fresh styles dark livid  . . . . . . . .
      . . . . . . . . . . . . . . . . . . . . . . . . . . . . . . . . . . . . . . . . . . . . . . . . . . . . . . . . . . . . . . . . . . . . . *H. subcrocatum*

6  Stem leaves 6 or 7 . . . . . . . . . . . . . . . . . . . . . . . . . . . . . . . . . . . . . . . . . . . . . . . . . . . . . . . . . . . . . . . .7
   Stem leaves 8 to 30 . . . . . . . . . . . . . . . . . . . . . . . . . . . . . . . . . . . . . . . . . . . . . . . . . . . . . . . . . . . . . . .9

7   Leaves yellowish or pale green, glabrescent above, inflorescence racemose-corymbose, phyllaries densely micro-glandular, with white pilose hairs ..............................*H. uiginskyense*
Leaves greyish green or glaucous, pilose or setose above, inflorescence furcate- or paniculate-corymbose, phyllaries with longer glandular hairs and dark-based pilose hairs ...................8

8   Stems with dense deflexed hairs at base, stem leaves elliptic or lanceolate, capitulum over 40 mm, phyllaries obtuse, floccose, with few glandular hairs, ligule tips glabrous, fresh styles yellowish or fuscous ........................................................*H. scoticum*
Stems pilose throughout, stem leaves ovate, capitulum 31–40 mm, phyllaries acute, sparingly floccose, with many glandular hairs, ligule tips pilose, fresh styles livid ................*H. iricum*

9   Phyllaries eglandular, or with few glandular hairs .........................................10
Phyllaries densely glandular ...........................................................11

10  Leaves glabrescent below, stem leaves not blotched, denticulate or distantly dentate, phyllaries floccose below only, densely pilose with white hairs and densely micro-glandular, fresh styles pure yellow ................................................................*H. uiginskyense*
Leaves pilose below, stem leaves often blotched, sub-entire or distantly denticulate, phyllaries efloccose, densely pilose with dark-based hairs, glands longer, fresh styles fuscous ..............
.........................................................................*H. sparsifolium*

11  Phyllary glandular hairs unequal, leaves lingulate or oblong-lanneolate, subentire or distantly denticulate.......................................................*H. strictiforme*
Phyllary glandular hairs equal, leaves elliptic-lanceolate, denticulate or dentate .................12

12  Leaves pilose below, bright or dark green, stem leaves acuminate, capitulum 31–40 mm, phyllary pilose hairs dark based, fresh styles fuscous or livid, receptacle pits with long bristles ..........
........................................................:..................*H. eboracense*
Leaves glabrescent below, yellowish or pale green, stem leaves acute, capitulum over 40 mm, phyllary pilose hairs white, fresh styles pure yellow, receptacle pits dentate-fimbriate..........
.........................................................................*H. uiginskyense*

13  Phyllaries eglandular, or with few glandular hairs .........................................14
Phyllaries densely glandular ...........................................................20

14  Radical leaves cuneate to attenuate below..................................................15
Radical leaves truncate or rounded at base, or abruptly narrowed at base .........................17

15  Peduncles with few glandular hairs, leaves yellowish green, long lanceolate, sparingly pilose above, peduncles suberect, capitulum 31–40 mm, phyllaries floccose..............*H. langwellense*
Peduncles with many dark glandular hairs, leaves dull or dark green or glaucous, ovate-lanceolate or ovate, more pilose above, peduncles incurved, capitulum over 40 mm, phyllaries sparingly floccose ...............................................................................16

16  Leaves dull or dark green, stem leaf petioles absent or short, stem leaves dentate at base, peduncles short, phyllaries not senescent, ligules yellow or deep yellow, fresh styles pure yellow ..............................................................*H. rubiginosum*
Leaves glaucous, stem leaf petioles long, stem leaves denticulate, peduncles long, phyllaries senescent, ligules light yellow, fresh styles livid ...............................*H. anglicum*

17  Heads geminate, leaves dull or dark green, phyllaries few and broad, dark or blackish green ......
.................................................................................*H. euprepes*
Heads not geminate, leaves bright green or glaucous, phyllaries numerous and narrow, greyish or olive green........................................................................18

18  Peduncles with few glandular hairs, peduncles sparingly floccose, phyllaries slightly senescent ...
.................................................................................*H. uisticola*
Peduncles with many white glandular hairs, or with many dark glandular hairs, peduncles floccose, phyllaries moderately or strongly senescent .....................................19

19  Peduncles with many pilose hairs, inner phyllaries incumbent in bud, phyllaries obtuse, margins densely floccose ......................................................*H. caesiomurorum*
Peduncles with few pilose hairs, inner phyllaries porrect in bud, phyllaries acute, margins floccose ............................................................*H. jovimontis*

20  Basal leaves truncate- or rounded based .....................................................21
Basal leaves cuneate based ...............................................................25

21  Phyllaries with few pilose hairs, fresh styles livid ..........................................22
Phyllaries with many pilose hairs, fresh styles pure yellow or fuscous ...........................23

22  Basal leaves obtuse, stem leaves lanceolate, leaves pilose but not floccose below, capitulum 20–30 mm, ligules orange-yellow . . . . . . . . . . . . . . . . . . . . . . . . . . . . . . . . . . . . . . . . . . . . . . . . . . *H. uistense*
    Basal leaves acute, stem leaves ovate, leaves glabrescent and floccose below, capitulum 31–40 mm, ligules light yellow . . . . . . . . . . . . . . . . . . . . . . . . . . . . . . . . . . . . . . . . . . . . . . . *H. shoolbredii*

23  Phyllaries densely floccose, leaves bright green, phyllaries greyish green, fresh styles fuscous . . . .
    . . . . . . . . . . . . . . . . . . . . . . . . . . . . . . . . . . . . . . . . . . . . . . . . . . . . . . . . . . . . . . . . . . . .*H. rivale*
    Phyllaries sparingly floccose or floccose below only, leaves glaucous, phyllaries dark or blackish green, fresh styles yellow . . . . . . . . . . . . . . . . . . . . . . . . . . . . . . . . . . . . . . . . . . . . . . . . . .24

24  Leaves more or less glabrous above, bright green and often purple-tinged, capitulum 31–40 mm  . .
    . . . . . . . . . . . . . . . . . . . . . . . . . . . . . . . . . . . . . . . . . . . . . . . . . . . . . . . . . . . . . . *H. pictorum*
    Leaves pilose or setose above, glaucous, capitulum from 31 mm to over 40 mm . . . . . . . . . . . . . . . . .59

25  Capitulum under 20 mm, inflorescence paniculate-corymbose . . . . . . . . . . . . . . . . . . . . . *H. duriceps*
    Capitulum from 20 to over 40 mm, inflorescence racemose or furcate-corymbose . . . . . . . . . . . . . . . .26

26  Phyllaries densely floccose, buds narrowed below, phyllaries greyish green . . . . . . . . . . . . . . .*H. rivale*
    Phyllaries efloccose or sparingly floccose, buds rounded below, or subtruncate, phyllaries dark or blackish green . . . . . . . . . . . . . . . . . . . . . . . . . . . . . . . . . . . . . . . . . . . . . . . . . . . . . . . . . . .27

27  Capitulum over 40 mm across, leaves setose below, leaf margins setose . . . . . . . . . . . . . . . . *H. schmidtii*
    Capitulum 20–40 mm, leaves glabrescent to pilose below, margins pilose or ciliate . . . . . . . . . . . . . . .28

28  Acladium very short, heads sometimes geminate, peduncles floccose, with fairly many glandular hairs . . . . . . . . . . . . . . . . . . . . . . . . . . . . . . . . . . . . . . . . . . . . . . . . . . . . . . . . . . *H. vennicontium*
    Acladium longer, heads not geminate, peduncles sparingly floccose, peduncles with many dark glandular hairs . . . . . . . . . . . . . . . . . . . . . . . . . . . . . . . . . . . . . . . . . . . . . . . . . . . . . . . . . . . .29

29  Leaves pale green or glaucous, peduncles suberect, phyllaries acute . . . . . . . . . . . . . . . .*H. hebridense*
    Leaves bright green, peduncles straight or incurved, phyllaries obtuse . . . . . . . . . . . . . . . . . . . . . . . .30

30  Leaves pilose below, peduncles straight, phyllaries sparingly floccose, dark green and with dark based pilose hairs, fresh styles pure yellow . . . . . . . . . . . . . . . . . . . . . . . . . . . . . . . . . . *H. nitidum*
    Leaves glabrescent or shortly pilose below, peduncles incurved, phyllaries efloccose, blackish green and with dark pilose hairs, fresh styles livid . . . . . . . . . . . . . . . . . . . . . . . . . . . . . . . .*H. sinuans*

31  Phyllaries densely glandular . . . . . . . . . . . . . . . . . . . . . . . . . . . . . . . . . . . . . . . . . . . . . . . . . . . .32
    Phyllaries eglandular or with few glandular hairs . . . . . . . . . . . . . . . . . . . . . . . . . . . . . . . . . . . . . .45

32  Capitulum under 20 mm across, inflorescence paniculate-corymbose . . . . . . . . . . . . . . . . . .*H. duriceps*
    Capitulum over 40 mm, inflorescence racemose-corymbose . . . . . . . . . . . . . . . . . . . . . . . . . . . . . . .33
    Capitulum 20–40 mm, inflorescence racemose- or furcate-corymbose . . . . . . . . . . . . . . . . . . . . . . . . .34

33  Stem leaves 5, leaves yellowish or pale green, peduncles medium, peduncles eglandular, bud truncate, phyllaries obtuse and with white pilose hairs . . . . . . . . . . . . . . . . . . . . . .*H. uiginskyense*
    Stem leaves 2 or 3, leaves glaucous, peduncles long and with many white glandular hairs, buds rounded below, phyllaries acute and with dark based pilose hairs . . . . . . . . . . . . . . . . . . *H. schmidtii*

34  Leaves shortly or moderately pilose above . . . . . . . . . . . . . . . . . . . . . . . . . . . . . . . . . . . . . . . . . . .35
    Leaves glabrous or glabrescent above . . . . . . . . . . . . . . . . . . . . . . . . . . . . . . . . . . . . . . . . . . . . . . .40

35  Acladium 1–2 mm, heads sometimes geminate, leaves shortly pilose above, glabrescent or shortly pilose below . . . . . . . . . . . . . . . . . . . . . . . . . . . . . . . . . . . . . . . . . . . . . . . . . . . *H. vennicontium*
    Acladium longer, heads not geminate, leaves pilose above and below . . . . . . . . . . . . . . . . . . . . . . .36

36  Peduncles with few glandular hairs . . . . . . . . . . . . . . . . . . . . . . . . . . . . . . . . . . . . . . . . . . . . . . . .37
    Peduncles with many white or dark glandular hairs . . . . . . . . . . . . . . . . . . . . . . . . . . . . . . . . . . . .38

37  Stem with 1 normal and 1 bractlike leaf, leaves yellowish or dull green, basal leaves rounded at base, phyllary margins densely floccose, ligules orange-yellow . . . . . . . . . . . . . . . . . . *H. caledonicum*
    Stem leaves 3–9, glaucous, basal leaves cuneate based, phyllary margins sparingly floccose, ligules full yellow . . . . . . . . . . . . . . . . . . . . . . . . . . . . . . . . . . . . . . . . . . . . . . . . . . . . . . . . . . . .*H. iricum*

38  Phyllaries densely floccose, leaves bright green, phyllary margins densely floccose . . . . . . . . . .*H. rivale*
    Phyllaries sparingly floccose, leaves yellowish, pale or dull green or glaucous, phyllary margins sparingly floccose . . . . . . . . . . . . . . . . . . . . . . . . . . . . . . . . . . . . . . . . . . . . . . . . . . . . . . . . . . .39

39  Basal leaves sharply or serrate-dentate, sometimes spotted, leaves rather small or moderate, fresh styles yellow or yellowish . . . . . . . . . . . . . . . . . . . . . . . . . . . . . . . . . . . . . . . . . . . . . *H. subrude*
    Basal leaves denticulate, not spotted, leaves rather large, fresh styles livid . . . . . . . . . . . .*H. hebridense*

40    Acladium 1–2 mm, heads sometimes geminate, buds subtruncate below ........... **H. vennicontium**
      Acladium longer, heads not geminate, buds narrowed or rounded or truncate below .............41

41    Peduncles with few glandular hairs, phyllaries dark green ...................................42
      Peduncles with many glandular hairs, phyllaries greyish orblackish green .......................43

42    Leaves moderately long, glabrescent above, yellowish or dull green, usually purplish below, outer
      phyllaries spreading, inner phyllaries with pale margins, acute, senescent, ligules orange
      yellow, fresh styles fuscous .................................................. **H. caledonicum**
      Leaves long, glabrous above, glaucous, paler below, outer phyllaries not spreading, inner phyllary
      margins not pale, obtuse, not senescent, ligules yellow, fresh styles yellow .......... **H. argenteum**

43    Leaves floccose above and below, dull green or glaucous ......................... **H. flocculosum**
      Leaves not floccose above or below, bright green .............................................44

44    Leaves pilose below, phyllaries greyish-green, densely floccose, with dark based pilose hairs .....
      .......................................................................................**H. rivale**
      Leaves glabrescent or short pilose below, phyllaries blackish-green, efloccose but with dark pilose
      hairs ................................................................................**H. sinuans**

45    Capitulum under 20 mm across, often stylose and geminate ...................... **H. cravoniense**
      Capitulum over 40 mm across ..........................................................46
      Capitulum 20–40 mm ..................................................................49

46    Peduncles with many dark glandular hairs and many pilose hairs.............................47
      Peduncles eglandular or with few glandular hairs and few pilose hairs .......................48

47    Leaves dull or dark green, stem leaf petioles absent or short, stem leaves dentate at base,
      peduncles short, phyllaries not senescent, ligules yellow or deep yellow, fresh styles pure
      yellow.................................................................... **H. rubiginosum**
      Leaves glaucous, stem leaf petioles long, stem leaves denticulate, peduncles long, phyllaries
      senescent, ligules light yellow, fresh styles livid ................................. **H. anglicum**

48    Leaves yellowish or pale green, glabrescent above and below, phyllaries densely pilose with white
      hairs and densely micro-glandular ....................................**H. uiginskyense**
      Leaves greyish green or glaucous, pilose or setose above, pilose below, phyllaries with many dark-
      based pilose hairs and longer glandular hairs....................................**H. scoticum**

49    Peduncles with many glandular hairs .......................................................50
      Peduncles eglandular or with few glandular hairs...........................................51

50    Leaves bright green or glaucous, phyllary margins densely floccose and tips markedly senescent ..
      ..............................................................................**H. caesiomurorum**
      Leaves yellowish, pale or dull green, sometimes spotted, phyllary margins sparingly floccose, tips
      not senescent ................................................................ **H. subrude**

51    Fresh styles pure yellow or yellowish .....................................................52
      Fresh styles fuscous or livid .............................................................55

52    Leaves glabrescent above, peduncles sparingly floccose ........................... **H. uisticola**
      Leaves pilose or with rough subsetiform hairs above, peduncles floccose .......................53

53    Basal leaves long lanceolate with long petioles, sparingly pilose above, stem leaves dentate at base,
      peduncles long, phyllaries floccose ......................................... **H. langwellense**
      Basal leaves elliptic, oblong, ovate or lanceolate, roughly pilose above and with short petioles,
      stem leaves denticulate, peduncles short or medium, phyllaries sparingly floccose..............54

54    Heads geminate, stem leaves 2, basal leaves denticulate or sinuate-dentate, acladium very short,
      peduncles with many pilose hairs, ligules mid to deep yellow, receptacle pits strongly
      fimbriate ................................................................... **H. euprepes**
      Heads not geminate, stem leaves 3 to 5, basal leaves distantly glandular denticulate, acladium of
      medium length, peduncles with few pilose hairs, ligules orange-yellow, receptacle pits
      dentate................................................................... **H. orimeles**

55    Leaves glabrous above ......................................................... **H. scarpicum**
      Leaves glabrescent or pilose above................................................56

56    Leaves long lanceolate, glabrescent above, peduncles with many pilose hairs, phyllaries olive
      green ................................................................**H. langwellense**
      Leaves ovate-lanceolate, oblong or lanceolate, pilose above, peduncles with few pilose hairs,
      phyllaries greyish or dark or blackish green .....................................57

57    Outer phyllaries spreading, inner linear-oblong, ligules orange-yellow . . . . . . . . . . . . . . *H. caledonicum*
      Outer phyllaries not spreading, inner subulate or linear-lanceolate, ligules mid-yellow . . . . . . . . . . . .58

58    Stem leaves contracted at base, usually with sharp ascending teeth, inflorescence paniculate-
      corymbose, inner phyllaries incumbent in bud, acuminate . . . . . . . . . . . . . . . . . . . . . . . . *H. vulgatum*
      Stem leaves attenuate at base, usually denticulate or with small, blunt teeth, inflorescence
      racemosecorymbose, inner phyllaries porrect in bud, acute . . . . . . . . . . . . . . . . . . . . . . *H. orimeles*

59    Leaves glabrescent or setose above, margins setose, acladium moderate, phyllaries moderately
      long, acute, with equal glandular hairs, capitulum over 40 mm . . . . . . . . . . . . . . . . . . . . . *H. schmidtii*
      Leaves pilose above, margins pilose, acladium short, peduncles with many dark glandular hairs,
      phyllaries long, attenuate at tip, with unequal glandular hairs, capitulum 31–40 mm . . . . . . . . . .
      . . . . . . . . . . . . . . . . . . . . . . . . . . . . . . . . . . . . . . . . . . . . . . . . . . . . . . . . . . . . . . .*H. ebudicum*

Subgenus **Hieracium**

Section **Subalpina** Pugsley

**H. sinuans** F. J. Hanb.
Rare.
One old record only, Barra, *Somerville*, 1888
(BM).   21

H. vennicontium Pugsl. is recorded in literature only, for Tarbert (*Shoolbred*) in Pugsley
(1948) and Lingadale (Harrison et al. 1942a),
both in North Harris.

Section **Cerinthoidea** Koch

**H. shoolbredii** E. S. Marshall
South Uist, Allt Volagir, *Pankhurst* 1979 (BM)
and Hecla, *Warburg* 1947 (BM); North Uist,
Eaval, *Wilmott* 1937 (BM); South Harris,
Beesdale, *West* 1955 (MNE); North Harris,
Strone Scourst, *J. W. Campbell* 1938 (BM),
Gillaval Glas, *Shoolbred* 1894 (BM).   9, 12,
16, 19

**H. flocculosum** Backh.
Rather similar to, but scarcer than the last.
South Harris, Luskentyre, *West* 1955 (CGE)
and Beesdale (ibid., MNE); Lewis, Stornoway
Castle, *Pankhurst* 1980 (BM); unconfirmed
records from South Uist by *J. W. H. Harrison*.
5, 12, (19)

**H. hebridense** Pugsley
This species is relatively frequent in Lewis
(Uig) and Harris. The collection by *M. S.
Campbell* at Ard Meavaig (Lewis) in 1937 is
the type (BM). South Uist (Allt Volagir),
Scarp and Harris, and Lewis (Uig).   3, 4, 9,
12, 19

**H. langwellense** F. J. Hanb.
Rare. One record only, Uisgnaval Beg,
North Harris (*Pankhurst*, 1984).   9.

**H. anglicum** Fries
First recorded by Balfour and Babington
(1841) from North Harris under the name *H.
murorum* var. *lawsoni*. Relatively common
and widely distributed. There are records for
var. *amplexicaule*.   3, 8, 9, 12, 16, 19, 21

**H. iricum** Fries
Relatively frequent in Harris and Lewis, but
absent from the southern islands.   3, 4, 8, 9,
12

**H. scarpicum** Pugsley
Endemic to the Outer Hebrides, and only
known from Scarp (hence the name), Lewis
and North Harris. The type specimens were
collected by *W. S. Duncan*, 1894, (BM).   3, 4,
8, 9

Section **Oreadea** Zahn

H. stenolepis (i.e H. stenolepiforme Pugsl.)
is recorded from Ben Eaval in North Uist by
*Shoolbred* in 1894 (Shoolbred, 1895), but no
herbarium material has been seen and the
record has not been confirmed. This species
is a rare endemic known only from Cheddar
in Somerset, and is unlikely to occur in the
Outer Hebrides.

**H. ebudicum** Pugsley
This species is endemic to the Outer Hebrides.
North Uist, Eaval, *Shoolbred* 1894 (NMW);
South Harris, Gilaval Glas, *Shoolbred* 1894
(NMW), Beesdale, *West* 1955 (MNE); North
Harris, Scaladale, *Wilmott* 1937 (BM), east of
Clisham, *Wilmott* 1939 (BM), Skeaudale
River, *Wilmott* 1939 (BM), Sgaoth Ard *J. W.
Campbell*, 1939 (BM); Lewis, Lingadale, *West*,
1955 (CGE), Cracaval *Bangerter* 1939 (BM).
The type is from Clisham in North Harris
(*Shoolbred*, 1895, BM).   3, 9, 12, 16

**H. schmidtii** Tausch
One of the commoner species, recorded from Barra to Lewis. Var. *crinigerum* is also recorded.    3, 4, 9, 10, 12, 18, 21

**H. nitidum** Backh.
Widespread but infrequent. Barra, 1936 (BM); South Harris, Beesdale, *West* 1955 (MNE); Taransay, Sythe harbour, Clark (1939c) det. *Pugsley*; North Harris, Clisham, *Shoolbred* 1894 (BM); Scarp, *W. S. Duncan* 1894 (BM).    3, 4, 9, 11, 12, 21

**H. jovimontis** (Zahn) Roffey
Rare. Recorded only from Barra, 1888 (BM) and from Coire Dubh, South Uist, (Harrison, 1941a), not confirmed.    (19), 21

**H. argenteum** Fries
Possibly the commonest and most wide-spread species. Here as elsewhere in Britain it is one of the relatively few species which can grow away from calcareous soils. There are records for var. *septentrionale* and var. *subglabratum*.    2–4, 8, 9, 12, 16, 21

**H. caledonicum** F. J. Hanb.
One of the commoner and more widespread species, known from Sandray to Lewis. Also widespread on the Scottish mainland.    3, 9, 12, 16, 18, 19, 21, 23

**H. scoticum** F. J. Hanb.
Like the last, relatively common and wide-spread, from Barra to Lewis.    3, 4, 9, 12, 16, 18, 21

**H. subrude** (Arvet-Touvet) Arvet-Touvet
Rare, Harris only. From Rodel in South Harris and Lingadale and the Maarig River in North Harris, all collected by *West* in 1955 (CGE, MNE).    9, 12

**H. orimeles** F. J. Hanb. ex W. R. Linton
(*H. beebyanum* Pugsl.)
Widespread and relatively frequent from Barra to Lewis.    3, 8, 9, 10, 12, 16, 19, 21

Section **Vulgata** F. N. Williams

**H. uistense** (Pugsley) P. D. Sell & C. West
Rare. The type locality is Ben Eaval, North Uist, where it was first collected by *Shoolbred* in 1894. The only collection from any other locality is from Valtos Glen, *Warburg* (BM), but hawkweeds now appear to have died out in that area.    3, 16

**H. duriceps** F. J. Hanb.
Rare by comparison with other parts of Scotland. Two records only, from Glen Beasdale in South Harris, *Miller* 1895 (BM) and an unconfirmed record by Harrison (1941a) from South Uist.    12, (19)

**H. pictorum** E. F. Linton
Rare, one record only in 1984 by *Pankhurst* and *Bevan* (BM) from river ravine of Gil Meodal, Horgabost, South Harris.    12

**H. rivale** F. J. Hanb.
Rare, one record only by Shoolbred (1899), from the Lee hills, North Uist. Needs confirmation.    (16)

**H. uisticola** Pugsley
Also rare and first recorded from the Lee Hills by *Shoolbred*, 1898, Pugsley (1948) which is the type locality (BM). There is also a single record from Mealisval, Lewis by *Bangerter and M. S. Campbell*, 1939 (BM), Pugsley (1942).    3, 16

**H. caesiomurorum** Lindeb.
Rare and only recorded from Allt Volagir in South Uist by *W. A. Clark* in 1942 (BM).    19

**H. euprepes** F. J. Hanb.
(*H. orcadense* W. R. Linton)
Scarce. South Uist, Allt Volagir, *Murray*, 1968 (CGE); South Harris, Tarbert, *Shoolbred* (1895) at NMW and MNE, and Gleann Shranndabhal, *Pankhurst*, 1979 (BM); and reported from Fuday by Harrison (1941a).    12, 19, (21)

**H. rubiginosum** F. J. Hanb.
(*H. orarium* Lindberg)
Relatively frequent. First recorded by *Shoolbred* from cliffs at Tarbert, South Harris in 1894 (Pugsley, 1948), where it still grows beside the pier for the ferry.    7, 9, 12, 16, 18, 19

**H. vulgatum** Fries
Rather scarce, by comparison with the Scottish mainland, where it is common on roadsides. No herbarium specimens have been seen.    16, 18, 19, 21

**H. cravoniense** (F. J. Hanb.) Roffey
Much less common than on the mainland, and only recorded from West Loch Tarbert, *M. S. Campbell*, 1938 (BM) and Rodel, *Wilmott and M. S. Campbell*, 1937 (BM) in South

Harris and once from North Harris, *Wilmott*, Pugsley (1948).   9, 12

Section **Tridentata** F. N. Williams

**H. sparsifolium** P. D. Sell & C. West
South Uist, Allt Volagir, *Wilmott*, 1947 (BM), Hecla, *J. A. Crabbe*, 1947 (BM); South Harris, Luskentyre, *Wilmott* 1939 (BM); Lewis, between Grimersta and Garrynahine, *M. S. Campbell*, 1939 (BM), Stornoway Castle, *Warburg*, 1946 (BM). There is an unlocalised specimen from North Harris, *W. S. Duncan* at BM.   3, 5, (9), 12, 19

**H. uiginskyense** Pugsley
Rare, known only from the north shore of Loch Boisdale, *M. S. Campbell*, 1947 (BM) and from Allt Volagir in South Uist, *Wilmott*, 1947 (BM).   19

**H. eboracense** Pugsley
Rare. Reported from Allt Volagir by *J. H. Harrison* without a voucher specimen.   (19)

Section **Foliosa** Pugsley

**H. latobrigorum** (Zahn) Roffey
Rare. Reported from Fuday by *J. H. Harrison* without a voucher specimen.   (21)

**H. subcrocatum** (E. F. Linton) Roffey
Widespread. Unconfirmed literature records from Barra, Eriskay, Scotasay and North Uist (Pugsley, 1948). South Harris, Drinishader, *M. S. Campbell*, 1939 (BM), Grosebay, *Pankhurst and Chater*, 1980 (BM); Scarp, *W. S. Duncan*, 1891 (BM); Lewis, Gravir, *Pankhurst and Chater*, 1980 (BM), Stornoway Castle (ibid.), near Barvas, *Fox* 1959 (BON).   1/2, 5, 7, 8, (10), 12, (16, 20, 21)

**H. strictiforme** (Zahn) Roffey
Fairly frequent. South Uist, Loch Skiport pier, *Wilmott*, 1947 (BM); North Uist, Eaval, *Wilmott*, 1937 (BM); Scarp, *West*, 1955 (MNE); Lewis, Stornoway Castle, *Warburg*, 1946 (BM), Uig, Kneep, *Pankhurst and Chater*, 1980 (BM), Geiraha, *M. S. Campbell*, 1939 (BM). 1, 3, 5, 8, 16, 19

**H. maritimum** (F. J. Hanb.) F. J. Hanb.
Rare. Collected by *Wilmott* from Carloway in Lewis in 1937. Although the specimen is only in bud, the species is very distinctive

and the record is confirmed by *Pugsley* (BM).   2

Subgenus **Pilosella** (Hill) S. F. Gray

Section **Pilosella** Fries

**H. pilosella** L.
(*Pilosella officinarum* C. H. & F. W. Schultz)
Mouse-ear Hawkweed.
Srubhan na Muice.
Native.
Machair, maritime heath and rocky moorland. Uncommon, mostly in the south. The first record was by Babington from the Shiant Isles, 1841, and this specimen at CGE has been determined as **subsp. tricholepium** by *Sell*.   3, 11–13, 15, 16, 19–22, 24, 27, 29
Three subspecies are recorded for the Outer Hebrides, based on specimens at BM determined by *Sell*. A key to all subspecies is given in Rich and Rich (1988).
subsp. **pilosella** at Rodel, South Harris.   12
subsp. **tricholepium** Naegeli & Peter. Mingulay to Lewis and the Shiant Isles.   3, 11, 15, 21, 24, 27
subsp. **melanops** Peter. Barra to South Harris.   12, 13, 16, 21

**H. aurantiacum** L.
(*Pilosella aurantiaca* (L.) C. H. & F. W. Schultz)
Fox-and-cubs.
Garden escape.
Recorded only once in North Uist (Harrison et al., 1942a).   16

**Crepis** L.
**C. capillaris** (L.) Wallr.
Smooth Hawk's-beard.
Lus Curain Mìn.
Native.
Waste places meadows and moorland.
J. B. Marshall has separated several variants of rather low status which we do not follow in this treatment.
Benbecula to Lewis.   1–3, 9, 12, 14–16, 18, 19, 21, 22

**Taraxacum** Wiggers
**Taraxacum officinale** Wiggers
Dandelion.
Beàrnan Brìde.
Native.
As a weed of waste places, native in machair and sand dunes, and on damp rocks on mountains. Widespread but surprisingly scarce.   2–5, 7, 9, 11–24, 28, 29

The following account of the microspecies of Taraxacum is based on information provided by A. J. Richards and C. C. Haworth. The nomenclature and numbering is correct according to a list circulated on 1st January 1986. Only records which are supported by herbarium specimens which have been determined by Richards, Haworth or van Soest are included. For accurate determination in Britain in general, the accepted ruling is that the specimens must have been collected before the end of May, except for alpine habitats. After the first flush of spring flowering, the plants lose their specific characters and take on what is known as 'summer form'. In the Outer Hebrides, this ruling is difficult to apply, since many plants do not begin to flower until June or July, even at sea level. Also, many plants are so poorly developed that a root or seed must be taken and plants grown in a garden before accurate determination is possible. The taxonomy of the genus has changed a great deal in recent years, and so many earlier records can only be allocated to the aggregate species, as above. The species present in the Outer Hebrides are those characteristic of north Britain and sand dune or mountain habitats, together with weedy species. There is none of the extensive endemism which occurs in the Shetlands (Scott and Palmer, 1987). The following list of microspecies is probably incomplete. More or less any of the weedy species of Sections Hamata and Vulgaria could occur as casuals.

In the key, the capitulum size is the diameter of the head when in flower.

## Key to *Taraxacum* in Outer Hebrides

1  Leaf lateral lobes none to 4 . . . . . . . . . . . . . . . . . . . . . . . . . . . . . . . . . . . . . . . . . . . . . . . . . . . .2
   Leaf lateral lobes 3 to 6 or more . . . . . . . . . . . . . . . . . . . . . . . . . . . . . . . . . . . . . . . . . . . . . .10

2  Ligule stripe carmine . . . . . . . . . . . . . . . . . . . . . . . . . . . . . . . . . . . . . . . . . . . . . . . . . . . . . . . .3
   Ligule stripe not carmine . . . . . . . . . . . . . . . . . . . . . . . . . . . . . . . . . . . . . . . . . . . . . . . . . . . . .5

3  Exterior bracts width 3–4 mm, capitulum 41–50 mm . . . . . . . . . . . . . . . . . . . . . . . . . . . . . *eximium*
   Exterior bracts width 2–3 mm, capitulum 21–40 mm . . . . . . . . . . . . . . . . . . . . . . . . . . . . . . . . .4

4  Upper surface of exterior bract not pruinose, exterior bracts unbordered, or scarcely bordered  . . . .
   . . . . . . . . . . . . . . . . . . . . . . . . . . . . . . . . . . . . . . . . . . . . . . . . . . . . . . . . . . . . . . . . . . . *faeroense*
   Upper surface of exterior bract pruinose, exterior bracts bordered . . . . . . . . . . . . . . . . . . . . *unguilobum*

5  Exterior bracts length 5–7 mm, scape slender, capitulum 21–30 mm, achene 2.6–3.0 mm . . . . *fulviforme*
   Exterior bracts length 7–12, scape medium, capitulum 31–50 mm, achene 3.1–4 mm . . . . . . . . . . . . . . .6

6  Upper (inner) surface of exterior bract purplish . . . . . . . . . . . . . . . . . . . . . . . . . . . . . . . . . . . . .7
   Upper (inner) surface of exterior bract green  . . . . . . . . . . . . . . . . . . . . . . . . . . . . . . . . . . . . . . .8

7  Distal margin of leaf lateral lobes entire, or denticulate, plant robust, leaf terminal lobe triangular, exterior bracts length 9–11 mm, upper surface of exterior bract not pruinose, exterior bracts unbordered . . . . . . . . . . . . . . . . . . . . . . . . . . . . . . . . . . . . . . . . . . . . . . . . . . . . . . . . . *atactum*
   Distal margin of leaf lateral lobes dentate, plant delicate, leaf terminal lobe sagittate, or helmet-shaped, exterior bracts length 7–9 mm, upper surface of exterior bract pruinose, exterior bracts scarcely bordered . . . . . . . . . . . . . . . . . . . . . . . . . . . . . . . . . . . . . . . . . . . . . . . . . . . . *fulvicarpum*

8  Upper surface of exterior bract pruinose, capitulum deep yellow  . . . . . . . . . . . . . . . . . . . . . . *praestans*
   Upper surface of exterior bract not pruinose, capitulum yellow . . . . . . . . . . . . . . . . . . . . . . . . . . . .9

9  Distal margin of leaf lateral lobes straight, or convex, exterior bracts length 9–11 mm, pollen present . . . . . . . . . . . . . . . . . . . . . . . . . . . . . . . . . . . . . . . . . . . . . . . . . . . . . . . . . . . . . *atactum*
   Distal margin of leaf lateral lobes concave, or sigmoid, exterior bracts length 7–9 mm, pollen absent . . . . . . . . . . . . . . . . . . . . . . . . . . . . . . . . . . . . . . . . . . . . . . . . . . . . . . . . . . . *maculosum*

10  Exterior bracts erect, or adpressed . . . . . . . . . . . . . . . . . . . . . . . . . . . . . . . . . . . . . . . . . . . . .11
    Exterior bracts recurved, or spreading . . . . . . . . . . . . . . . . . . . . . . . . . . . . . . . . . . . . . . . . . . .29

11  Exterior bracts width 3–6 mm . . . . . . . . . . . . . . . . . . . . . . . . . . . . . . . . . . . . . . . . . . . . . . . . .12
    Exterior bracts width up to 3 mm . . . . . . . . . . . . . . . . . . . . . . . . . . . . . . . . . . . . . . . . . . . . . . .13

12  Capitulum 0–20 mm, leaves smooth, scape slender, exterior bracts length up to 5 mm, exterior bracts corniculate, ligule involute . . . . . . . . . . . . . . . . . . . . . . . . . . . . . . . . . . . . . . . . . *obliquum*
    Capitulum 31–60 mm, leaves neutral texture or rough, scape medium, or broad, exterior bracts length 7–14 mm, exterior bracts smooth, ligule flat . . . . . . . . . . . . . . . . . . . . . . . . . . . . . . . .14

13  Petiole winged . . . . . . . . . . . . . . . . . . . . . . . . . . . . . . . . . . . . . . . . . . . . . . . . . . . . . . . . . . . .18
    Petiole unwinged . . . . . . . . . . . . . . . . . . . . . . . . . . . . . . . . . . . . . . . . . . . . . . . . . . . . . . . . . . .24

14  Pollen absent, upper surface of exterior bracts not pruinose . . . . . . . . . . . . . . . . . . . . . . . . . . . . . . . . .15
    Pollen present, upper surface of exterior bracts pruinose  . . . . . . . . . . . . . . . . . . . . . . . . . . . . . . . . . .16

15  Leaf lateral lobes linear, or narrow, or medium width, leaves lanceolate, exterior bracts erect,
      capitulum 31–40 mm, ligule stripe not carmine . . . . . . . . . . . . . . . . . . . . . . . . . . . . . . . . . . *maculosum*
    Leaf lateral lobes broad, leaves spathulate, exterior bracts adpressed, capitulum 41–50 mm, ligule
      stripe carmine . . . . . . . . . . . . . . . . . . . . . . . . . . . . . . . . . . . . . . . . . . . . . . . . . . . . . . . . . . . . . . . . . *eximium*

16  Distal margin of leaf lateral lobes entire, or denticulate, capitulum deep yellow . . . . . . . . . . . *raunkiaerii*
    Distal margin of leaf lateral lobes dentate, capitulum yellow. . . . . . . . . . . . . . . . . . . . . . . . . . . . . . .17

17  Leaves spotted, leaf shade dark, exterior bracts length 11–14 mm, capitulum 51–60 mm . . . . *naevosum*
    Leaves unmarked, leaf shade light, exterior bracts length 7–11 mm, capitulum 31–40 mm . . . . *laetifrons*

18  Plant robust. . . . . . . . . . . . . . . . . . . . . . . . . . . . . . . . . . . . . . . . . . . . . . . . . . . . . . . . . . . . . . . . . . . . . . .19
    Plant delicate. . . . . . . . . . . . . . . . . . . . . . . . . . . . . . . . . . . . . . . . . . . . . . . . . . . . . . . . . . . . . . . . . . . . . . .21

19  Leaf lateral lobes patent . . . . . . . . . . . . . . . . . . . . . . . . . . . . . . . . . . . . . . . . . . . . . . . . . . . . . . . . *raunkiaerii*
    Leaf lateral lobes recurved . . . . . . . . . . . . . . . . . . . . . . . . . . . . . . . . . . . . . . . . . . . . . . . . . . . . . . . . . . .20

20  Distal margin of leaf lateral lobes entire, or denticulate, leaf shade dark, upper surface of exterior
      bracts not pruinose, lower (outer) surface of exterior bracts blackish, pollen absent . . . . . . *spectabile*
    Distal margin of leaf lateral lobes dentate, leaf shade light, upper surface of exterior bracts
      pruinose, lower (outer) surface of exterior bracts green, pollen present . . . . . . . . . . . . . . . . .*laetifrons*

21  Distal margin of leaf lateral lobes entire . . . . . . . . . . . . . . . . . . . . . . . . . . . . . . . . . . . . . . . . . . *spectabile*
    Distal margin of leaf lateral lobes denticulate, or dentate . . . . . . . . . . . . . . . . . . . . . . . . . . . . . . . .22

22  Exterior bracts bordered, leaf shade light . . . . . . . . . . . . . . . . . . . . . . . . . . . . . . . . . . . . . *platyglossum*
    Exterior bracts unbordered, or scarcely bordered, leaf shade medium, or dark. . . . . . . . . . . . . . . . .23

23  Midrib pink to purple, distal margin of leaf lateral lobes denticulate, ligule stripe carmine. . . *spectabile*
    Midrib white to green, distal margin of leaf lateral lobes dentate, ligule stripe not carmine . . *proximum*

24  Ligule stripe carmine, lower (outer) surface of exterior bracts blackish . . . . . . . . . . . . . . . . . . *spectabile*
    Ligule stripe not carmine, lower (outer) surface of exterior bracts green. . . . . . . . . . . . . . . . . . . . . .25

25  Exterior bracts length 3–7 mm, exterior bracts corniculate, capitulum 21–30 mm . . . . . . . . . . . . *laetum*
    Exterior bracts length 7–10 mm, exterior bracts smooth, capitulum 31–40 mm . . . . . . . . . . . . . . . . .26

26  Pollen present. . . . . . . . . . . . . . . . . . . . . . . . . . . . . . . . . . . . . . . . . . . . . . . . . . . . . . . . . . . . . . . . . . . . .27
    Pollen absent. . . . . . . . . . . . . . . . . . . . . . . . . . . . . . . . . . . . . . . . . . . . . . . . . . . . . . . . . . . . . . . . . . . . . . .28

27  Plant robust, leaf lateral lobes medium width, or broad, upper surface of exterior bracts pruinose. .
      . . . . . . . . . . . . . . . . . . . . . . . . . . . . . . . . . . . . . . . . . . . . . . . . . . . . . . . . . . . . . . . . . . . . . . . . . . *raunkiaerii*
    Plant delicate, leaf lateral lobes linear, or narrow, upper surface of exterior bracts not pruinose . . . .
      . . . . . . . . . . . . . . . . . . . . . . . . . . . . . . . . . . . . . . . . . . . . . . . . . . . . . . . . . . . . . . . . . . . . . . . . . . *landmarkii*

28  Achene spinulose to halfway, achene 3.6–4 mm, achene cone 0.4–0.6 mm . . . . . . . . . . . . . . *maculosum*
    Achene spinulose above, achene 3.1–3.5 mm, achene cone 0–0.3 mm . . . . . . . . . . . . . . . . . . *landmarkii*

29  Petiole unwinged . . . . . . . . . . . . . . . . . . . . . . . . . . . . . . . . . . . . . . . . . . . . . . . . . . . . . . . . . . . . . . . . . . .30
    Petiole winged . . . . . . . . . . . . . . . . . . . . . . . . . . . . . . . . . . . . . . . . . . . . . . . . . . . . . . . . . . . . . . . . . . . . . .39

30  Upper (inner) surface of exterior bract purplish . . . . . . . . . . . . . . . . . . . . . . . . . . . . . . . . . . . . . . . . .31
    Upper (inner) surface of exterior bract green . . . . . . . . . . . . . . . . . . . . . . . . . . . . . . . . . . . . . . . . . . . . .32

31  Exterior bracts corniculate, leaf lateral lobes linear, or narrow, leaf terminal lobe sagittate, or
      tripartite, scape slender . . . . . . . . . . . . . . . . . . . . . . . . . . . . . . . . . . . . . . . . . . . . . . . . . *brachyglossum*
    Exterior bracts smooth, leaf lateral lobes medium width, or broad, leaf terminal lobe triangular, or
      helmet-shaped, scape medium . . . . . . . . . . . . . . . . . . . . . . . . . . . . . . . . . . . . . . . . . . . . . . . . . . . . . . . .33

32  Leaves spotted . . . . . . . . . . . . . . . . . . . . . . . . . . . . . . . . . . . . . . . . . . . . . . . . . . . . . . . . . . . . . . . . . . . . . .35
    Leaves unmarked. . . . . . . . . . . . . . . . . . . . . . . . . . . . . . . . . . . . . . . . . . . . . . . . . . . . . . . . . . . . . . . . . . . .36

33  Leaves spotted . . . . . . . . . . . . . . . . . . . . . . . . . . . . . . . . . . . . . . . . . . . . . . . . . . . . . . . . . . . . . . *naevosiforme*
    Leaves unmarked, or blotched . . . . . . . . . . . . . . . . . . . . . . . . . . . . . . . . . . . . . . . . . . . . . . . . . . . . . . . . .34

34  Leaf lateral lobes patent . . . . . . . . . . . . . . . . . . . . . . . . . . . . . . . . . . . . . . . . . . . . . . . . . . . . . . . . *raunkiaerii*
    Leaf lateral lobes recurved . . . . . . . . . . . . . . . . . . . . . . . . . . . . . . . . . . . . . . . . . . . . . . . . . . . . . .*subhamatum*

35  Capitulum deep yellow, achene spinulose throughout, achene 3.1–3.5 mm . . . . . . . . . . . . *naevosiforme*
    Capitulum yellow, achene spinulose in lower part only, achene 3.6–4 mm . . . . . . . . . . . . . . *maculosum*

Section Erythrosperma (Lindb.f.) Dahlst.

**T. brachyglossum** (Dahlst.) Raunk.
Native.
South Uist, Rubha Ardvule, *Wilmott* 1939 (BM).   19

**T. laetum** Dahlst.
Native.
North Uist, Baleshare dunes, *McAllister* 1979 (BM).   16

**T. fulviforme** Dahlst.
Native.
North Uist, Kirkibost and Loch Portain, both by *J. W. Clark*, 1980 (E).   16

**T. fulvum** Raunk.
Native.

North Uist, Kyles Paible, *J. W. Clark* 1980 (E); Harris, Ensay, *M. S. Campbell*, 1939 (BM); Lewis, Valtos, *J. A. Campbell*, 1939 (BM).   3, 14, 16

**T. proximum** (Dahlst.) Raunk.
Native.
Benbecula, Nunton, *Pankhurst*, 1979 (BM); South Harris, Grosebay, *Pankhurst*, 1979 (BM).   12, 18

Section Obliqua Dahlst.

**T. obliquum** (E.Fries) Dahlst.
Native.
South Harris, Luskentyre. The first British record, *Wilmott* (BM), 1946.   12.

**T. platyglossum** Raunk.
Native.
North Uist, dunes on Baleshare, *J. W. Clark*, 1989 (E).   16

Section Spectabilia Dahlst.

Some authorities regard the next three taxa as conspecific.

**T. eximium** Dahlst.
Native.
South Uist, Beinn Mhor, *Warburg* (BM).   19

**T. faeroense** Dahlst.
Native.
Mountain gullies, wet rocks. The most frequent species.
Lewis, Cracaval; North and South Harris; North Uist, Sidinish; South Uist, Ben Corodale.   3, 9, 12, 16, 19

**T. spectabile** Dahlst.
Native.
Similar habitats to the last.
South Uist, Hecla, *Warburg* (BM); North Uist, Loch Portain, *J. W. Clark*, 1979 (E); North Harris, Cravadale, *Pankhurst*, 1979 (BM).   9, 16, 19

Section Naevosa M. P. Christiansen

**T. euryphyllum** (Dahlst.) Christ.
Native.
North Uist, Baleshare, *J. W. Clark*, 1980 (E); South Harris, Rodel, *M. S. Campbell*, 1939 (BM).   12, 16

**T. fulvicarpum** Dahlst.
Native.
North Uist, Newton, *Wilmott*, 1937 (BM); South Harris, Northton, *Wilmott*, 1937 (BM); Lewis, Europie Bay, *Wilmott* 1946 (BM) and Uig, *Wilmott*, 1937 (BM).   1, 3, 12, 16

**T. maculosum** A. J. Richards (= *maculigerum* sensu A. J. Richards)
Native.
North Uist, Newton Ferry, *J. W. Clark*, 1980 (E); Harris, Tarbert, *Cameron*, 1979 (BM).   9, 16

**T. naevosiforme** Dahlst.
Native.
Barra, Cuier, *Wilmott* (BM).   21

**T. naevosum** Dahlst.
Native.

Barra, Cuier, *Wilmott* (BM); Berneray (Harris), *M. S. Campbell* 1939 (BM); South Harris, Rodel, *M. S. Campbell* 1939 (BM); Lewis, Mangersta, *Crabbe and Bangerter*, 1939 (BM).   3, 12, 15, 21

Section Celtica A. J. Richards

**T. unguilobum** Dahlst.
Native.
Surprisingly rare, compared with Scotland generally.
North Uist, Carinish, *J. A. Clark* 1980 (E).   16

**T. laetifrons**
Native.
North Uist, Sidinish, *J. W. Clark* 1980 (E); Shillay, *M. S. Campbell* 1939 (BM).   13, 16

**T. landmarkii** Dahlst.
Native.
North Uist, Bayhead, *J. W. Clark*, 1980 (E).   16

**T. praestans** H. Lindb. f.
Native.
North Uist, Lochmaddy, *J. W. Clark* S23/1980 (E) det. *Richards* but det. as **T. raunkiaeri** by *Haworth* 1988. Nevertheless a common species in Scotland and likely to occur.

**T. raunkiaeri** Wiinst.
Native.
North Uist, Lochmaddy, *J. W. Clark*, 1980 (E).   16

Section Hamata H. Øllgaard

**T. atactum** Sahlin and van Soest
Native.
North Uist, Kirkibost, *J. W. Clark* S17/1980 (E) det. *Richards* but indet. by *Haworth* 1988. A common species in Scotland and likely to occur.   16

**T. subhamatum** Christ.
Native.
North Uist, Bayhead, *J. W. Clark*, 1980 (E).   16

Section Ruderalia Kirschner, Øllgaard & Štěpánek
(Section *Vulgaria* Dahlstedt)

**T. alatum** H. Lindb. f.
Native
North Uist, Bayhead and Lochmaddy, both *J. W. Clark*, 1980 (E).   16

**T. subcyanolepis** Christ.
Native.

North Uist, Balmore, *J. W. Clark*, 1980 (E).
16

LILIOPSIDA
ALISMATALES
*ALISMATACEAE*
**Baldellia** Parl.
**B. ranunculoides** (L.) Parl.
Lesser Water-plantain.
Corr-chopag Bheag.
Native.
Mesotrophic loch margins and streams.
Barra to North Uist and Monach Islands. 16,
18, 19, 21, 29

NAJADALES
*JUNCAGINACEAE*
**Triglochin** L.
**T. palustris** L.
Marsh Arrowgrass.
Bàrr a' Mhilltich Lèana.
Native.
Marshes, flushes, ditches and saltmarsh.
Mingulay to Lewis. 1–5, 7, 9, 10, 12, 13,
15–24, 27

**T. maritima** L.
Sea Arrowgrass.
Bàrr a' Mhilltich Mara.
Native.
Saltmarsh.
Mingulay to Lewis and Monach Is. 1–5, 7,
12, 13, 15–19, 21, 22, 24, 29

*ZOSTERACEAE*
**Zostera** L.
**Z. marina** L.
Eelgrass.
Bilearach.
Native.
Maritime mudflats.
Vatersay to Sound of Harris and Loch Roag,
Lewis. 3, 14–16, 18, 19, 20, 21, 22

**Z. angustifolia** (Hornem.) Reichenb.
Narrow-leaved Eelgrass.
Bilearach na Duilleige Caoile.
Native.
Mudflats.
First recorded (as *Z. marina* var. *angustifolia*)
Loch Stromore, North Uist *Shoolbred* 1894
(BM, E); Alioter, North Uist *J. W. Clark* 1978
(det. *Matthews*); Loch Bru, near Alioter,
North Uist *Chater* and *Chorley* 1983 (BM). 16

**Z. noltii** Hornem.
*Z. nana* auct.

Recorded in error in BSBI Atlas (fide Preston).
This same error has been copied over into
the National Vegetation Classification,
which shows a dot on the map for the
community based on this species.

*POTAMOGETONACEAE*
**Potamogeton** L. (by C. D. Preston)
Pondweed.
The Outer Hebrides have a particularly
rich Potamogeton flora, more than two-
thirds of the British species occurring in the
archipelago. These are concentrated in the
relatively eutrophic and calcareous lochs on
the landward fringe of the machair from
Barra to North Uist. The occurrence here in
proximity of so many species has led to the
presence of a number of hybrids, some local
or even rare. Elsewhere in the islands the
predominantly oligotrophic waters hold few
Potamogeton species, although the presence
of **P. epihydrus** in such habitats on South
Uist is of outstanding phytogeographical
interest.
Credit for the discovery of this rich
Potamogeton flora must go to J. W. H.
Harrison and his team, who were the first to
collect **P. alpinus, P. × billupsii, P. coloratus, P. epihydrus, P. praelongus, P. × prussicus, P. pusillus, P. × suecicus, P. × sparganifolius** and **P. rutilus**. Their observations
are summarized by J. W. H. Harrison (1949).
Heslop Harrison's earlier specimens were
seen by the late J. E. Dandy and G. Taylor,
the foremost authorities on the genus. Unfortunately a dispute between *J. W. Heslop
Harrison* on the one hand and *Dandy* and *G.
Taylor* on the other, developed into a bitter
wrangle and led to perhaps the most vitriolic
printed exchanges in British field botany this
century (Dandy and Taylor 1942, 1944;
Harrison 1944; Harrison and Clark 1941b,
1942a). Harrison (1944) refused to allow
Dandy and Taylor access to any of his later
collections, and voucher specimens of **P. ×
cooperi, P. × gessnacensis, P. × heslop-
harrisonii, P. lucens** and **P. obtusifolius**
cannot now be traced. In their absence these
records cannot be accepted, and are
included in square brackets in the list below.
Since 1950 the unpolluted, species-rich
lochs on South Uist, Benbecula and North
Uist have attracted an increasing number of
botanists, and their Potamogeton flora is
now relatively well-known. Particularly significant visits were made by G. Taylor (in
1951), Miss U. K. Duncan (in 1960) and,

especially, by Mrs. *J. W. Clark*, whose collecting in the early 1970s was encouraged by *J. E. Dandy*. Mention should also be made of surveys by the Royal Botanic Garden, Edinburgh (in 1983 and 1984) and the ecological study published by Spence et al. (1979). Much less attention has been paid to other aquatic habitats, and to the less species-rich lochs elsewhere in the Outer Hebrides.

Potamogetons are often misidentified, and literature records should not be accepted uncritically. For rarer taxa, all records accepted below are unless stated, based on specimens determined by Dandy or Dandy and Taylor, or, in the case of material they did not see, by *C. D. Preston* or *N. F. Stewart*. *J. E. Dandy* kept a comprehensive card index of Potamogeton records, which is now at the BM. 'Dandy Index' indicates a specimen seen by Dandy and Taylor and recorded on Dandy's card index, but the current whereabouts of which is not known. In the case of rare taxa collected more than twice from the same locality, only the first and the latest records are cited.

**P. natans** L.
Broad-leaved Pondweed.
Duileasg na h-Aibhne, Lìobhag.
Native.
Lochs, lochans, streams and ditches; in oligotrophic and acidic to eutrophic and alkaline waters. Its ecological requirements overlap with those of **P. polygonifolius** and in the Hebrides both species often grow in the same oligotrophic lochs and lochans.
Sandray to Lewis, frequent. 1–5, 7–9, 11, 12, 14–21, 23, 29

[*P.* × *gessnacensis* G. Fisch. (*P. natans* L. × *P. polygonifolius* Pourret) was reported by J. W. H. Harrison (1949) from Loch na Liana Moire, Benbecula.]

**P. polygonifolius** Pourret
Bog Pondweed.
Lìobhag Bogaich.
Native.
Oligotrophic lochs and lochans, streams, flooded peat cuttings, wet streamsides and flushes; calcifuge.
Berneray to Lewis; the most frequent Potamogeton species away from the machair. 1–10, 12–24, 27–29

**P. coloratus** Hornem.
Fen Pondweed.
Native.

Base-rich shallow fen pools and machair lochs.
Monach Islands; Loch Sniogravat, *W. A. Clark* 1938 (Dandy Index); *Perring* 1949 (CGE). Benbecula; pool near Creagorry, *J. W. H. Harrison* 1940 (Dandy Index); Loch na Liana Moire, *W. A. Clark* and *J. W. H. Harrison* 1940 (Dandy Index); *U. K. Duncan* 1960 (Dandy Index); shallow fen pools north of Loch na Liana Moire, *Preston, Stewart* et al. 1987 (CGE). 18, 29
These are the northernmost localities in Great Britain for this predominantly eastern, calcicole species.

**P.** × **billupsii** Fryer
(= *P. coloratus* Hornem. × *P. gramineus* L.)
Machair loch.
Benbecula; Loch na Liana Moire, *J. W. H. Harrison* 1940 (E); *Preston, Stewart* et al. 1987 (CGE); loch on roadside near Borve Castle, *J. W. H. Harrison* 1940 (Dandy Index).
Loch na Liana Moire is now the only known site for this hybrid, which has long been extinct at its type locality in Cambridgeshire. 18

[*P. lucens* L. MacGillivray's (1830) record was made when the critical study of the British Potamogeton species had scarcely begun, and should be disregarded. A record from Barra (Watson and Barlow 1936; Campbell 1936) is based upon a specimen later determined as **P. gramineus** (Dandy and Taylor 1940). Heslop Harrison (in Vasculum 28 (3): 24, 1943; see also Harrison 1949) reported **P. lucens** 'with certainty' from a swift runnel near Loch Kildonan, South Uist. No voucher specimen has been seen and there is no subsequent proof of the occurrence of this large and conspicuous species. A reference to extensive beds of **P. lucens** in the moor lochs of North Uist (Waterston et al. 1979) is almost certainly erroneous.]

**P. gramineus** L.
Various-leaved Pondweed.
Native.
Lochs, more rarely streams and ditches; tolerant of a wide range of alkalinity and salinity (Spence et al. 1979) but most frequent in the machair lochs. Sandray to South Harris and probably to Lewis. 1–3, 7, 8, 12, 15, 16, 18, 19, 21, 23

**P.** × **sparganifolius** Laest. ex Fr.
(= *P. gramineus* L. × *P. natans* L.)

Native.
Benbecula; Loch na Liana Moire, *W. A. Clark* and *J. W. H. Harrison* 1940 (Dandy Index); *U. K. Duncan* 1960 (BM).   18
When discovered in Loch na Liana Moire **P. × sparganifolius** covered considerable areas of the loch, sometimes growing with one or both parents. It spread vegetatively, but never flowered (Harrison 1949). No specimen has been seen to support the record from Loch Cistavat, South Harris (Harrison and Clark 1941b, Harrison et al. 1941a).

**P. × nitens** Weber
(= *P. gramineus* L. × *P. perfoliatus* L.)
Lochs, streams and large ditches.
Barra to North Lewis.    1, 2, 8, 12, 15, 16, 18, 19, 21
The most frequent Potamogeton hybrid in the archipelago.

[*P. × heslop-harrisonii* W. A. Clark. The most mysterious Potamogeton in the British flora, described by Clark (1943) from Loch Grogary, North Uist, and later reported from Loch Mhor, Baleshare (Harrison and Harrison 1950a). There are no later records. Clark (1942) initially reported it as **P. gramineus** × **P. berchtoldii** but after detailed study decided that it was **P. gramineus** × **P. perfoliatus** × **P. berchtoldii**. Heslop Harrison (1949) suggested that the parentage **P. alpinus** × **P. berchtoldii** could not be ruled out. Dandy and Taylor never saw any specimens of **P. × heslop-harrisonii** and recent enquiries have failed to reveal any. Only a few hybrids between broad-leaved and narrow-leaved species of Potamogeton are known, and all are rare.]

**P. alpinus** Balbis
Red Pondweed.
Native.
Lochs, lochans and drainage ditches, rare.
Benbecula, North Uist and North Lewis.
Lewis; runnel from Loch Stiapavat to Ness, *J. W. H. Harrison* 1949 (K, cf Harrison and Harrison 1950b). Benbecula; lochans near Nunton and Oban Uaine, *J. W. H. Harrison* 1940 (Dandy Index). South Uist; Loch na Clacha-mora, *W. A. Clark* 1946 (BM); *G. Taylor* 1951 (BM); Loch an Eilean, *G. Taylor* 1951 (BM); drainage ditch, Loch Hallan, *Chamberlain* 1983 (E).
There are literature records from a ditch by Loch na Tanga, South Uist, 1984 (RBG, Edinburgh, 1984) and Loch Leodasay (Nicol

1936). Nicol's record is dubious as the plant has not been seen at Loch Leodasay by subsequent botanists.

**P. × prussicus** Hagstr.
(*P. × johannis* Heslop Harrison
= *P. alpinus* Balbis × *P. perfoliatus* L.)
With **P. alpinus** in lochan near Oban Uaine, Benbecula *J. W. H. Harrison* 1940 (Dandy Index, cf Dandy and Taylor 1941 and Harrison and Clark 1941a).
Neither **P. × prussicus** nor either of its parents have been rediscovered in recent searches of the lochs and lochans near Oban Uaine.

**P. praelongus** Wulf.
Long-stalked Pondweed.
Native.
South Uist.   19
In deep water (over 1.5 m) in large, meso-trophic lochs.
South Uist; Loch Kearsinish, *J. H. Harrison* 1938 (Dandy Index); Loch Druidibeg, *Williams* 1958 (BM); *Bowen* 1969 (BM); East Loch Ollay, *Chorley and Chater* 1983 (BM); Mid Loch Ollay, *Preston, Stewart* et al., 1987 (CGE). In addition to these localities, there are literature records from Loch Ceann a' Bhaigh, Loch Fada and Loch Teanga, South Uist, 1977 (Spence et al. 1979). A record from Barra (Watson and Barlow 1936; Campbell 1936) is an error for **P. gramineus** (Dandy and Taylor 1940).

**P. perfoliatus** L.
Perfoliate Pondweed.
Native.
Lochs, most frequent in machair lochs but occasionally found in more oligotrophic waters.
Barra to South Harris and probably Lewis. 2, 3 or 5, 12, 15–19, 21

**P. epihydrus** Raf. var. **ramosus** (Peck) House
American Pondweed.
Native.
Lochs and peaty lochans.
South Uist.   19
South Uist; Loch Ceann a' Bhaigh, *J. W. H. Harrison* 1943 (E, cf Harrison 1949, 1950b, 1952); lochans between L. Ceann a' Bhaigh and L. Ollay, *W. A. Clark* 1944 (BM, cf Harrison 1951a); *Preston, Stewart* et al. 1987 (CGE); Loch an Duin, *W. A. Clark* 1945 (Dandy Index); *U. K. Duncan* 1960 (E).

These are the only native localities in Europe for this species, which is widespread in North America. It was first found in Loch Ceann a' Bhaigh and in a single nearby lochan. Since 1983 it has been seen in five peaty lochans between Loch Ceann a' Bhaigh and Loch Ollay, some of which may be flooded peat cuttings. Here it flowers and fruits freely in the oligotrophic water (conductivity 95–131 µS cm$^{-1}$ at 25 C in September 1987). Other species occurring at these sites include *Eleocharis multicaulis, Eleogiton fluitans, Juncus bulbosus* var. *fluitans, Lobelia dortmanna, Menyanthes trifoliata, Nymphaea alba, Potamogeton natans, P. polygonifolius* and *Sparganium angustifolium*. These lochans are apparently no different to many hundreds of others in the Hebrides and the very restricted distribution of **P. epihydrus** suggests that it may have arrived, by natural means, relatively recently. It would be interesting to know if subfossil fruits can be found in the loch sediments of South Uist. All the material from South Uist is referable to var. **ramosus** (Peck) House (var. *nuttallii* (Cham. & Schlecht.) Fernald), one of two varieties of **P. epihydrus** recognized in North America by Fernald (1932) and the predominant one in the eastern part of its range.

This species is listed in the Red Data Book, Perring and Farrell (1983).

**P. friesii** Rupr.
Flat-stalked Pondweed.
Native.
Machair lochs.
Barra to Berneray and Monach Is.    15, 16, 18, 19, 21, 29
Berneray (Harris); Little Loch Borve, *J. W. Clark* 1972 (BM). North Uist; Loch Mor, Baleshare *J. W. Clark* 1973 (BM); *Preston, Stewart* et al. 1987 (CGE). Benbecula; Loch Uacraich, *McKean* et al., 1984 (E). South Uist; *Somerville* 1888 (CGE); Loch Hallan and Loch na Liana Mhoir, *McKean* 1983 (E); Loch Stilligary, *McKean* 1983 (E); Loch Grogarry, *Chamberlain* 1984 (E). Barra; Loch na Doirlinn, *Wilmott* 1938 (BM); *Bratton* 1987 (specimen discarded); Loch St. Clair, *M. S. Campbell* 1978 (BM), *Chater* 1983 (BM). Monach Islands; Loch nam Buard, *Young* 1969 (BM).

**P. rutilus** Wolfg.
Shetland Pondweed.
Native.
Machair lochs.

North Uist.    16
North Uist; Loch Grogary, *W. A. Clark* 1942 (Dandy Index); *Preston* and *Stewart,* 1987 (CGE); Loch Scarie, *W. A. Clark* 1942 (BM); *Preston* and *Stewart, W. A. Clark* 1942 (CGE); Loch Leodasay, *J. W. Clark* 1975 (E); Loch Mor, Baleshare, *J. W. H. Harrison* 1942 (BM); *David* 1961 (LTR).
A northerly species in the British Isles, recorded from Shetland, Tiree, E. Ross and Easterness as well as North Uist.
This species is listed in the Red Data Book, Perring and Farrell (1983).

**P. pusillus** L.
Lesser Pondweed.
Native.
Machair lochs and nearby streams and ditches.
South Uist to North Uist.    16, 18, 19
The most frequent linear-leaved species in the machair lochs. The record from Barra (Conacher 1980) is an error; the material (BM) is **P. friesii**. No specimens have been seen to support the records from South Harris (Harrison and Morton 1951), Scarp (Harrison 1949) and Loch Eilaster, Lewis (Harrison and Harrison 1950b).

[**P. obtusifolius** Mert. & Koch. A 'somewhat interesting form' of this species was reported from Loch Snigisclett, South Uist, by *J. W. H. Harrison* in Vasculum 28 (3): 24, 1943. He did not, however, include the record in his otherwise comprehensive paper on the Hebridean Potamogetons (Harrison 1949). He presumably decided that it had been a misidentification, perhaps for **P. berchtoldii** which is reported from Loch Snigisclett as subsp. **lacustris** Pearsall & Pearsall f. in the later paper.]

**P. berchtoldii** Fieb.
(*P. millardii* Heslop Harrison)
Small Pondweed.
Native.
Lochs, lochans and streams.
Barra to Lewis.    1–3, 8, 15, 16, 18, 19, 21, 29
The ecological differences between **P. berchtoldii** and the similar **P. pusillus** are well shown in the Outer Hebrides, where **P. berchtoldii** is less frequent in the machair lochs but occurs elsewhere in more oligotrophic and acidic waters.

**P. crispus** L.
Curled Pondweed.

Native.
Machair lochs and streams.
South Uist to Berneray and probably to Lewis, rather infrequent.   3, 15, 16, 18, 19, 29

[*P.* × cooperi (Fryer) Fryer (*P. crispus* L. × *P. perfoliatus* L.) was reported by Harrison et al. (1942b) from a runnel leading from a lochan near Ormaclett, South Uist.]

**P. filiformis** Pers.
Slender-leaved Pondweed.
Native.
Machair lochs, usually in shallow water over a sandy substrate.
Barra to Berneray and perhaps Lewis.   1, 12, 15, 16, 18, 19, 21, 29
A characteristic species of the seaward side of machair lochs, often in the **Chara aspera–Potamogeton filiformis** association (Spence et al. 1979).

**P.** × **suecicus** K. Richter
(= **P. filiformis** Pers. × **P. pectinatus** L. )
Native.
Benbecula and Berneray.
Benbecula; lochs near Borve Castle and Uachdar, *W. A. Clark* 1940 (Dandy Index); Loch na Liana Moire, *W. A. Clark* 1940 (Dandy Index); *Preston, Stewart* et al. 1987 (CGE). Berneray (Harris); Loch Bhruist, *Campbell* 1938 (Dandy Index); *Wilmott* 1939 (Dandy Index); Little Loch Borve, *W. A. Clark* 1939 (Dandy Index). Clark (1943) reported **P.** × **suecicus** from Baleshare, North Uist, noting that he could find no **P. filiformis** there. His voucher material (BM), however, is actually **P. filiformis**. No specimens have been seen to support the records from South Uist (Harrison and Clark 1942b, Harrison et al. 1942a, Harrison 1949) and Loch Cistavat, South Harris (Harrison and Harrison 1950b). **P.** × **suecicus** is easily overlooked and its occurrence in South Uist would not be surprising. Characters for separating the hybrid from its parents are given by Dandy and Taylor (1940).

**P. pectinatus** L.
Fennel Pondweed.
Native.
Machair lochs and other lochs, pools and streams by the sea; notably tolerant of brackish water.
Barra to North Lewis, locally frequent.   1–3, 9, 12, 14–16, 18, 19, 21, 29

The rhizomes, tubers, stems, leaves and seeds of this species are a significant food for wildfowl, especially dabbling duck. J. W. Campbell (1946) investigated the stomach contents of 156 widgeon (*Anas penelope*) shot in North Uist, and found 10 in which the chief constituent was **P. pectinatus**.

*RUPPIACEAE*
**Ruppia** L.
Largely based on material determined by J. E. Dandy.
**R. cirrhosa** (Petagna) Grande
*R. spiralis* L. ex Dumort.
Spiral Tassel-weed.
Native.
Brackish lochs.
Benbecula and the Uists.   16, 18, 19

**R. maritima** L.
Beaked Tasselweed.
Native.
Brackish water.
Barra to Lewis, less restricted than the last species.   1–3, 16, 18, 19, 21

*ZANNICHELLIACEAE*
**Zannichellia** L.
**Z. palustris** L.
Horned Pondweed.
Lìobhag Adhairceach.
Native.
Brackish and mesotrophic lochs.
South Uist to Lewis and Monach Is.   1, 16, 18, 19, 29

*NAJADACEAE*
**Najas** L.
**N. flexilis** (Willd.) Rostk. & Schmidt
Slender Naiad.
Aibhneag.*
Native.
An uncommon plant of a few lochs in the Uists, first discovered by Harrison (1941b, 1952).   16, 19
This is a protected species, listed in the Red Data Book, Perring and Farrell (1983).

LILIIFLORAE
*LILIACEAE*
**Narthecium** Hudson
**N. ossifragum** (L.) Hudson
Bog Asphodel.
Bliochan.
Native.
Damp moorland and flushes.

Berneray to Lewis, St Kilda to Shiant Is. 1–7, 9–13, 15–24, 27, 28

**Scilla** L.
**S. verna** Hudson
Spring Squill.
Lear-uinnean.
Native.
Cliff tops and maritime heaths.
Berneray and Mingulay to Butt of Lewis. 1–3, 13, 16, 19–24

**Hyacinthoides** Medicus
**H. non-scripta** (L.) Chouard ex Rothm.
(*Endymion non-scriptus* (L.) Garcke)
Wild Hyacinth [Bluebell in England]
Bròg na Cuthaig, Fuath-mhuc.
Native.
Sea and lochside cliffs and stream ravines.
Barra to southern Lewis. 7, 16–21

**Allium** L.
**A. ursinum** L.
Ramsons.
Creamh, Gairgean.
Native.
Cliffs and woods.
Scattered from Barra to Lewis. 5, 12, 16, 19, 21

*JUNCACEAE*
**Juncus** L.
**J. maritimus** Lam.
Sea Rush.
Meithean.
North Uist, from salt marsh, Carinish, Clark (1939c), Baleshare, *J. W. Clark* 1982; South Uist, Clark (1939c), Harrison (1941a) and Loch Eynort, Carnan and Lochboisdale (ibid.). 16, 19

**J. filiformis** L.
Thread Rush.
Native.
Very rare, one locality only; Loch Arnol, Lewis *Norman* 1984 (BM) det. *Cope.* 2
This is a protected species, listed in the Red Data Book, Perring and Farrell (1983).

**J. balticus** Willd.
Baltic Rush.
Luachair Bhailtigeach.
Native.
Damp machair.
South Uist to Lewis. 1–3, 12, 13, 15, 16, 18, 19

**J. inflexus** L.
Hard Rush.
Probably introduced.
Only one record; Barra, in a layby, *Conacher* (1980). 21

**J. effusus** L.
Soft-rush.
Luachair Bhog.
Native.
Damp pastures, moorland and marshes.
Berneray and Mingulay to Lewis and St Kilda. 1–7, 9, 12, 13, 15–24, 27–29

**J. conglomeratus** L.
*J. subuliflorus* Drejer
Compact Rush.
Brodh-bràighe.
Native.
On damp moors and marshes.
Rather scarce, but throughout the Islands. 1–7, 10–12, 16, 18–21, 23, 24

**J. trifidus** L.
Three-leaved Rush.
Luachair Thrì-bhileach
Native.
Mountain rocks.
Only one record; Ullaval, c. 1500 ft, North Harris *W. S. Duncan* 1896 (BM, E)

**J. squarrosus** L.
Heath Rush.
Brù-chorcan, Moran.
Native.
Moorland.
Mingulay to Lewis and St Kilda. 1, 3–7, 9–13, 16–22, 24, 27, 28

**J. compressus** Jacq.
Round-fruited Rush.
Native.
Only record; by stream, Gerinish, Loch Bee, South Uist *Poulter* 1951 (E). 19
Unlikely to occur in the Islands, possibly an error for the next species.

**J. gerardi** Lois.
Saltmarsh Rush.
Luachair Rèisg Ghoirt.
Native.
Saltmarsh.
Sandray to Lewis. 1–5, 7, 8, 10, 12, 14–23, 27, 29

**J. tenuis** Willd.
Slender Rush.

Luachair Chaol.
Colonist.
Tracks and paths.
Only record from Haskeir, Monach Islands
*Atkinson* 1952 (BM).   29

**J. bufonius** group
This treatment is based on specimens seen
by *Cope* and *Stace*
**J. bufonius** L.
Toad Rush.
Buabh-luachair.**
Native.
Loch margins and machair.
Mingulay to Lewis and St Kilda to Shiant
Is.   1–12, 14–24, 25, 27–29

**J. ambiguus** Guss.
Native.
Saline habitats and saltmarsh.
First record det. in 1966 by *Holub* North Rona
*Darling* 1939 (CGE); Mingulay Bay,
Mingulay *Chater* 1983 (BM); Loch Bee, South
Uist *Chater* 1983 (BM); Rubha Ardvule,
South Uist *Chater* 1983 (BM); W. of Cheese
Bay, North Uist *Chater* 1983 (BM)
Mingulay to North Uist and North Rona. 16,
19, 24, 25

J. capitatus Weigel
Dwarf Rush.
Recorded in SW Barra by *Clark* and *J. W. H.
Harrison* (Harrison, 1939 and 1941a) (BM, E,
K). This species has not subsequently been
found although searched for.

**J. bulbosus** L.
Bulbous Rush.
Luachair Bhalgach.
Native.
Lochs, streams and flushes.
Berneray and Mingulay to Lewis and St
Kilda.   1, 3–7, 9–22, 24, 27–29
The larger plants with six stamens are some-
times separated as *J. kochii* Schultz. The
aquatic var. *fluitans* is recorded.

**J. acutiflorus** Ehrh. ex Hoffm.
Sharp-flowered Rush.**
Luachair a' Bhlàth Ghèir.
Native.
Marshes and damp places.
Barra to Butt of Lewis.   1, 3, 4, 6, 7, 10, 12,
13, 16, 18–23

One plant from Luskentyre, South Harris
*Wilmott* and *M. S. Campbell* 1939 (BM)

appears to be **J.** × **surrejanus** Druce ex Stace
& Lambiñon (J. acutiflorus × articulatus)

**J. articulatus** L.
Jointed Rush.
Lachan nan Damh.
Native
Marshes, flushes and lochsides.
Berneray and Mingulay to Lewis, St Kilda to
North Rona.   1–4, 6, 7, 9, 11–19, 21–24, 27,
29

**Luzula** DC.
**L. campestris** (L.) DC.
Field Wood-rush.
Learman Raoin.
Native.
Heathy moorland and banks.
Berneray and Mingulay to Lewis, North
Rona to St Kilda.   1–3, 5, 7, 9, 12, 15–25, 28,
29

**L. multiflora** (Retz.) Lej.
Heath Wood-rush.
Learman Monaidh.
Native.
Moorland, marshes, damp pasture and
machair.
Berneray and Mingulay to Lewis and St
Kilda.   1–7, 9, 10, 12–19, 21, 22, 24, 27–29
Var. **congesta** (Thuill.) Hyl. has been occa-
sionally recorded.

**L. spicata** (L.) DC.
Spiked Wood-rush.
Learman Ailpeach.
Native.
Very rare on mountain tops in Harris;
Mullach an Langa, Ceartaval, Uisgnaval
More and Clisham on North Harris and
Roneval on South Harris.   9, 12

**L. sylvatica** (Hudson) Gaudin
Great Wood-rush.
Luachair Coille.
Native.
Stream ravines, mountain slopes, Damp
moorland and plantations.
Mingulay to Lewis and St Kilda.   2–5, 7,
9–12, 16–19, 21–24, 27, 28

**L. pilosa** (L.) Willd.
Hairy Wood-rush.
Learman Fionnach.
Native.
Heathy moorland and mountain ledges.
Quite rare, Barra to Harris.   12, 16, 18, 19, 21

*IRIDACEAE*
**Iris** L.
**I. versicolor** L.
Purple Iris.
Escape from cultivation.
South Harris, near Leverburgh, Harrison
(1957) and field record, with I. pseudacorus,
*J. W. Campbell* in 1959.   12
This is a protected species, listed in the Red
Data Book, Perring and Farrell (1983).

**I. pseudacorus** L.
Yellow Iris.
Seileasdair, Sealasdair.
Native.
Near houses, wet meadows and by streams.
Berneray and Mingulay to Lewis, St Kilda to
Shiant Is.    1–7, 9–24, 27–29

**Tritonia** Ker-Gawler
*Crocosmia* Planch. p.p.
**T.** × **crocosmiflora** (Lemoine) Nicholson
*T. aurea* Pappe ex Hooker × *pottsii* (Baker)
Baker
*Crocosmia* × *crocosmiflora* (Lemoine) N. E. Br.
Montbretia.
Hortal.
First collection, roadside between Clachan
and Carinish, North Uist *M. S. Campbell* 1972
(BM); field records from Barra, Berneray
(Harris) and the Eye penisula and literature
record (Harrison, 1957) from South Harris.
6, 12, 15, 16, 21

ORCHIDALES
*ORCHIDACEAE*
**Spiranthes** L. C. M. Richards
**S. romanzoffiana** Cham.
Irish Lady's-tresses
Mogairlean Bachlach Bàn.*
Native.
Marshy ground.
Very local but sometimes in quantity, first
record; Barra *Robarts* 1967 (E); Benbecula
*Begg* 1977; Vatersay *Chorley* 1983 (BM)
Vatersay, Barra and Benbecula.    18, 21, 22
This is a protected species, listed in the Red
Data Book, Perring and Farrell (1983).

**Listera** R. Br.
**L. ovata** (L.) R. Br.
Common Twayblade.
Dà-dhuilleach.
Native.
Machair and pasture.
Barra and Fuday to Lewis.    3, 9, 12, 15, 16,
19, 21, 22

**L. cordata** (L.) R. Br.
Lesser Twayblade.
Dà-dhuilleach Monaidh.
Native.
Damp moss beneath Heather.
Barra to Lewis.    1–5, 9, 11, 12, 16, 18, 19, 21

**Hammarbya** O. Kuntze
**H. paludosa** (L.) O. Kuntze
Bog Orchid.
Mogairlean Bogaich.
Native.
Wet bogs.
Vatersay to Lewis and St Kilda.    3, 7, 9, 12,
19, 21, 22, 28

**Coeloglossum** Hartmann
**C. viride** (L.) Hartmann
Frog Orchid.
Mogairlean Losgainn.
Native.
Machair.
Vatersay to Lewis, Monach Is. to St Kilda.
1–3, 8, 9, 12–16, 18, 19, 21, 22, 28, 29

Putative hybrids involving the above species
have been reported; only records with
voucher specimens have been considered.
× **Dactyloglossum** × **mixtum** (Ascherson &
Graebner) Rauschert (*Coeloglossum viride* ×
*Dactylorhiza maculata*). Both specimens seen
appear to belong here; Husinish, South
Harris *Raven* 1948 (BM); Fuday Is, Barra
*Ferreira* 1955 (K).    12, 21

J. H. Harrison (1949) described *Orchis viridella*
from Borve, Harris ascribing the parentage
*Coeloglossum viride* × *Dactylorhiza purpurella*
but no specimen has been found to support
the record.

**Gymnadenia** R. Br.
**G. conopsea** (L.) R. Br.
Fragrant Orchid.
Lus Taghte.
Native.
Machair.
Recorded only from Fuday, Barra, where it
is abundant.    21
Harrison (1941a:260) names it as subsp.
**insulicola**. He also described the hybrid
between **G. conopsea** subsp. **insulicola** ×
**Dactylorhiza** (Orchis) **fuchsii** var. **hebriden-
sis** (now known as × **Dactylogymnadenia
cooksii** Nelson) from Corodale Bay, Fuday
but no specimen can be found. Rich and
Rich (1988) mention subsp. **borealis** which is

smaller in its parts and has a northern distribution but there is no record from the Quter Isles.

Pseudorchis Séguier
P. albida (L.) R. Br.
Small-white Orchid.
Mogairlean Bàn Beag.
Recorded from Loch a' Chlachain, South Uist Harrison et al. (1939) and Harrison (1941a) but not confirmed.

Platanthera L. C. M. Richard
P. chlorantha (Custer) Reichenb.
Greater Butterfly-orchid.
Mogairlean an Dealain-dè Mòr.
Native.
Rare, in South Harris only; Obbe W. S. Duncan 1891 (BM) and Rodel Glen J. W. Campbell 1938 (BM).    12
P. bifolia (L.) L. C. M. Richard
Lesser Butterfly-orchid.
Mogairlean an Dealain-dè Beag.
Native.
Machair and moorland.
Barra to North Harris.    9, 11, 12, 16, 19, 21

Orchis L.
O. mascula L.
Early-purple Orchid.
Moth-ùrach.
Native.
Cliffs and woods.
Barra to Lewis.    3, 5, 6, 9, 11, 12, 16, 20, 21, 22
Small forms from Taransay, South Harris and Fuday were separated as var. ebudicum by J. W. H. Harrison.

Dactylorhiza Necker ex Nevski
Dactylorchis (Klinger) Vermeulen
D. incarnata (L.) Soó
subsp. incarnata
Early Marsh-orchid.
Mogairlean Lèana.
Native.
Marshy ground near the sea, machair.
Mingulay to Lewis.    3, 4, 9, 11–13, 15, 16, 18, 19–24, 29
subsp. coccinea (Pugsley) Soó
Native.
Machair and dunes.
Seems to have a more northern distribution in the Islands than the last subspecies.
Barra to Lewis.    1, 3, 9, 14, 16, 18, 19, 21

D. maculata (L.) Soó
subsp. ericetorum (E. F. Linton) P. F. Hunt & Summerhayes
Heath Spotted-orchid.
Mogairlean Mòintich.
Native.
Moorland, machair and damp pastures.
Berneray and Mingulay to Lewis and St Kilda.    1–7, 9, 11–24, 28, 29

D. fuchsii (Druce) Soó
subsp. hebridensis (Wilmott) Soó
Hebridean Spotted-orchid.
Urach-bhallach.
Native.
Machair and damp meadows especially coastal.
Mingulay to Lewis and St Kilda.    1–4, 8–10, 12, 15, 16, 18–24, 28
All the plants of the Islands seem to be this subspecies with a very few apparently intermediate with subsp. fuchsii.
Bateman and Denholm (1989) give a detailed description of the variation of this taxon in the Outer Hebrides and elsewhere. They view this and other subspecies of D. fuchsii as varieties.

D. × kerniorum (Soó) Soó
O. × variabilis, D. fuchsii subsp. hebridensis × incarnata
Barra, Harrison (1949); Eoropie, Lewis J. W. H. Harrison 1949 (K); several specimens from South Uist Davis 1951 (E)

D. × transiens (Druce) Soó
D. fuchsii subsp. hebridensis × maculata subsp. ericetorum
Vatersay to South Harris.    12, 19, 21, 22

D. × venusta (T. & T. A. Stephenson) Soó
O. hebridella Wilmott, D. fuchsii subsp. hebridensis × purpurella
Barra, Harrison (1949); Eoropie, Lewis J. W. H. Harrison 1949 (K); Gerinish, South Uist Davis 1951 (E); Eoligarry, Barra Ferreira 1955 (K).    1, 19, 21

D. × formosa (T. & T. A. Stephenson) Soó
D. maculata subsp. ericetorum × purpurella
A widely recorded hybrid.
Barra to Lewis.    1, 3, 4, 9, 12, 16, 18, 19, 21

J. H. Harrison recorded D. maculata subsp. rhoumensis (Heslop Harrison f.) Soó from Shillay but this is now considered restricted to Rhum, Inner Hebrides.

**D.** × **dingelensis** (Wilmott) Soó
*D. maculata* subsp. *ericetorum* × *majalis* subsp. *occidentalis*
SW of Clachan, North Uist *Wilmott* and *M. S. Campbell* 1934 (4 specimens, BM).   16

**D.** × **latirella** (P. M. Hall) Soó
*D. incarnata* × *purpurella*
Barra, Harrison (1949); Eoropie, Butt of Lewis *J. W. H. Harrison* 1949 (K) and many sites Barra to Lewis.   1, 2, 9, 11, 12, 16, 18, 21

**D. incarnata** subsp. **coccinea** × **majalis** subsp. **occidentalis**
recorded once from marsh SW of Clachan, North Uist *Wilmott* 1937 (BM)

**D. majalis** (Reichenb.) P. F. Hunt & Summerhayes
subsp. **occidentalis** (Pugsley) P. D. Sell
*D. kerryensis (Wilmott)* P. F. Hunt & Summerhayes
Broad-leaved Marsh-orchid.
Native.
Calcareous marshes and damp machair.
South Uist to Lewis.   1, 2, 12, 15, 16, 19

subsp. **purpurella** (T. & T. A. Stephenson) D. Moresby Moore & Soó
*D. purpurella* (T. & T. A. Stephenson) Soó
Northern Marsh-orchid.
Mogairlean Purpaidh.
Native.
Marshy fields, machair and higher ground.
Mingulay to Lewis.   1–6, 8, 9, 11, 12–16, 18, 19, 21, 23, 24, 29

Some experts consider **D. purpurella** to be a separate species from **D. majalis** and that the populations in the Outer Hebrides are mostly referable to **D. purpurella** subsp. **majaliformis** Nelson, with the exception of populations on the north coast of North Uist which agree with **D. majalis** subsp.**scotica** Nelson.

**D. majalis** subsp. **occidentalis** × subsp. **purpurella**
Hybrids of this parentage are reported from Northton, South Harris by Harrison (1951a), and there is a specimen from SW of Clachan, North Uist *Wilmott* 1937 (BM).

**D. lapponica** (Laest. ex Hartman) Soó
Native.
Basic hill-flushes.

Previously identified as **D. traunsteineri** (Sauter) Soó, see Kenneth, Lowe and Tennant (1988).
Rare, one small population in South Harris only.   12

**Anacamptis** L. C. M. Richard
**A. pyramidalis** (L.) L. C. M. Richard
Pyramidal Orchid
Mogairlean nan Coilleag.
Native.
Base-rich grassland and machair.
*J. W. H. Harrison* described var. **fudayensis** from Fuday, *W. A. Clark* in 1938(K), differs from the type only in having an elongate, rather cylindrical spike. In 1986, despite careful searching, *Turner-Etlinger* was unable to rediscover this taxon.
Barra (Eoligarry) and Fuday; Askernish and Daleburgh, South Uist.   19, 21

ARALES
*ARACEAE*
**Arum** L.
**A. maculatum** L.
Lords-and-ladies.
Cluas Chaoin.
Probably a garden escape.
Outside garden, Newton, North Uist *M. S. Campbell* 1936 (BM); same site *J. W. Campbell* 1938 (BM).   16

*LEMNACEAE*
**Lemna** L.
**L. minor** L.
Common Duckweed.
Mac gun Athair, Aran Tunnaig.
Ditches and pools.
Barra to North Uist, Berneray and the Monach Is.   15–19, 21, 29

TYPHALES
*SPARGANIACEAE*
**Sparganium** L.
**S. erectum** L.
(*S. ramosum* Hudson)
Branched Bur-reed.
Seisg Rìgh.
Native.
Lochs and ditches.
Rich and Rich (1988) give a key to four subspecies, for whose determination ripe fruit is essential. From what is know of the distribution of these subspecies, subsp. **microcarpum** (Neuman) Domin and subsp. **neglectum** (Beeby) Schinz and Thell. may be

expected in the Outer Hebrides. One specimen from Balranald Marshes, North Uist *Shoolbred* may represent subsp. **microcarpum**. Barra to North Uist, and one locality in Lewis (Loch Stiapavat).   1, 16, 19, 21

**S. emersum** Rehmann
Unbranched Bur-reed.
Seisg Madaidh.
Native.
Mesotrophic lochs.
*Wilmott* first recorded this from Loch by School 2 miles S of Carloway, Lewis 1937; Askernish, South Uist *M. S. Campbell* et al. 1947 (BM); Stilligarry *Pankhurst* 1983 (BM). 2, 19, ?21

**S. angustifolium** Michx
Floating Bur-reed.
Seisg air Bhog.
Native.
Oligotrophic lochs.
Barra to Lewis.  1–4, 6, 7, 9, 10, 12, 14, 16–19, 21

**S. affinis** L.
*S. minimum* Wallr.
Least Bur-reed.
Seisg as Lugha.
Native.
Oligotrophic lochs.
Barra to Lewis.  3, 16, 18, 19, 21

*TYPHACEAE*
**Typha** L.
**T. latifolia** L.
Bulrush.
Cuigeal nam Ban-sìdh.
Probably introduced.
South Uist, south end of Upper Loch Kildonan, *Cadbury* 1978; Lewis, reported from Loch Ordais by Harrison (1957), also from Loch Arnol and Loch Ordais, Biagi et al. (1985). 2, 19

CYPERALES
*CYPERACEAE*
**Eriophorum** L.
**E. angustifolium** Honckeny
Common Cottongrass.
Canach.
Native.
Bogs, flushes and damp moorland.
Berneray and Mingulay to Lewis, St Kilda to North Rona.  1–7, 9, 10, 12–25, 27, 28

**E. latifolium** Hoppe
Broad-leaved Cottongrass.

Canach na Duilleige Leathainn.
Native.
More base-rich places than the last species. Only recorded fron North of Loch Langavat, North Harris *M. S. Campbell* 1937 and *Wilmott* 1939 (both BM).   9

**E. vaginatum** L.
Hare's-tail Cottongrass.
Sìoda Monaidh.
Native.
Damp moorland and bogs.
Mingulay to Lewis.  1, 3–7, 9, 11, 12, 15, 16, 18, 19, 21, 22, 24

**Trichophorum** Pers.
**T. cespitosum** (L.) Hartman
subsp. **germanicum** (Palla) Hegi
Deergrass.
Cìob, Ultanaich.
Native.
Moorland.
There is apparently no record for subsp. **cespitosum**.
Berneray and Mingulay to Lewis and St Kilda.  1–7, 9, 11–24, 27, 28

**Eleocharis** R. Br.
**E. acicularis** (L.) Roemer & Schultes
Needle Spike-rush.
Reported from North Rona by Barrington (1885), but unlikely and never confirmed.

**E. quinqueflora** (F. X. Hartmann) O. Schwarz
(*E. pauciflora* (Lightf.) Link)
Few-flowered Spike-rush.
Bioran nan Lusan Gann.
Native.
Flushes.
Barra to Lewis and St Kilda.  1, 3, 7, 9, 12, 13, 15, 16, 18, 19, 21, 22

**E. multicaulis** (Sm.) Sm.
Many-stalked Spike-rush.
Bioran Badanach.
Native.
Flushes, damp moorland, lochs and streamsides.
Berneray to Lewis.  1–4, 7, 9, 10, 12, 13, 16–24

**E. palustris** (L.) Roemer & Schultes
Common Spike-rush.
Bioran Coitcheann.
Native.
Lochs, marshes and ditches.
Berneray and Mingulay to Lewis, St Kilda and North Rona.  1–7, 9, 12–25, 27–29

All the material appears to be referable to subsp. **vulgaris** Walters

**E. uniglumis** (Link) Schultes
Slender Spike-rush.
Bioran Caol.
Native.
Shallow water, often brackish.
Benbecula to Lewis and St Kilda.    1–3, 7, 9, 12–14, 16, 18, 19, 28

**Scirpus** L.
**S. maritimus** L.
Sea Club-rush.
Bròbh.
Native.
Ditches,    streams,    brackish    lochs    and saltmarsh.
Pabbay and Barra to Gt Bernera (rare in the north of the Islands).    4, 6, 12, 13, 16, 18, 19, 21, 23, 29

**Blysmus** Panzer
**B. compressus** (L.) Panzer
Flat-sedge.
Seisg Rèidh.
Has been recorded from Loch Hallan near Daliburgh, South Uist *J. W. H. Harrison* et al. (1939), and Harrison (1941a); the records may refer to forms of the next species.

**B. rufus** (Hudson) Link
Saltmarsh Flat-sedge.
Seisg Rèisg Ghoirt.
Native.
Saltmarsh and brackish loch margins.
Barra to Lewis.    1–4, 12, 15, 16, 18, 19, 21

**Schoenoplectus** (Reichenb.) Palla
**S. lacustris** L.
subsp. **lacustris**
(*Scirpus lacustris* L.)
Common Club-rush.
Luachair Ghòbhlach.
Native.
South Uist to Butt of Lewis (but rarer than next).    1, 12, 16, 19

subsp. **tabernaemontani**(C. C. Gmelin) A. & D.Löve
(*Scirpus tabernaemontani* C. C. Gmelin)
Grey Club-rush.
Luachair Bhogain.
Native.
Lochs, streams and ditches.
Barra to North Harris.    9, 12, 16, 18, 19, 21

**Isolepis** R. Br.
**I. setacea**  (L.) R. Br.
(*Scirpus setaceus* L.)
Bristle Club-rush.
Curcais Chalgach.
Native.
Loch and streamsides.
Sandray to Lewis.    1–3, 6, 7, 9, 12, 13, 15, 16, 18, 19, 21–23

**I. cernua** Vahl
(*Scirpus cernuus* Vahl)
Slender Club-rush.
Curcais Chaol.
Native.
Lochs and streamsides.
Mingulay and Barra.    21, 22, 24

**Eleogiton** Link
**E. fluitans** L.
*Scirpus fluitans* L.
Floating Club-rush.
Curcais air Bhog.
Native.
Oligotrophic lochs and streams.
Mingulay to Lewis.    1–5, 7, 9, 12, 15–24

**Schoenus** L.
**S. nigricans** L.
Black Bog-rush.
Sèimhean.
Native.
Maritime grassland and moorland flushes.
A prostrate form is characteristic of maritime grassland.
Berneray and Mingulay to Lewis and St Kilda.    1–5, 7, 9, 11–13, 15, 16, 18, 19, 21–24, 28

**Rhynchospora** Vahl
**R. alba** (L.) Vahl
White Beak-sedge.
Gob-sheisg.
Native.
Bogs.
Barra to Lewis.    3, 6, 7, 9, 12, 16, 18, 19, 21

**Cladium** Browne
**C. mariscus** (L.) Pohl
Great Fen-sedge.
Sàbh-sheisg.
Native.
Margins of oligotrophic lochans. The habitat in the Outer Hebrides is unusual for this species.
Scarce, only in Eriskay, Benbecula and Uists.    16, 18, 19, 20

**Carex** L.
**C. paniculata** L.
Greater Tussock-sedge.
Seisg Bhadanach Mhòr.
Native.
Marshes.
Benbecula and the Uists.   16, 18, 19

**C. diandra** Schrank
Lesser Tussock-sedge.
Seisg Bhadanach Bheag.
Native.
Machair, marshes and reed-beds.
Pabbay (Harris), Pitkin et al. (1983); North Uist, Balranald Marshes, *Shoolbred* 1898 (BM), Houghgarry, *Cadbury* 1978; South Uist, Loch na Tanga, *Warburg* 1947 (BM), Gerinish, *Davis* 1951 (E), frequent in machair lochs in the southern part of the island, *Cadbury* 1978; Shiant Isles.   13, 14, 16, 19, 21, 27

**C. disticha** Hudson
Brown Sedge.
Seisg Dhonn.
Native.
Fens.
South Uist, Loch a' Ghearraidh Dhuibh, Kilpheder, *Cadbury* 1978, and Loch Toronish, Chamberlain et al. (1984); the specimen from Barvas, Lewis *J. W. H. Harrison* 1949 (K) has been redetermined by *Mullin* as **C. arenaria** L. There is an unconfirmed record from Seilebost, South Harris, Harrison (1957). 12?, 19

**C. otrubae** Podp.
False Fox-sedge.
Seisg Gharbh Uaine.
Native.
Upper part of saltmarshes and rock crevices.
Pabbay and Barra to North Uist and Ensay to Shiant Isles.   14, 16, 18, 19, 21, 23, 27

**C. arenaria** L.
Sand Sedge.
Seisg Gainmhich.
Native.
Dunes and machair.
Mingulay to Lewis and Monach Isles to St Kilda.   1–3, 9, 11, 12, 14–16, 18–24, 28, 29

**C. maritima** Gunnerus
Curved Sedge.
Seisg Bheag Dhubh-cheannach.
Native.
Wet sand by the sea.
North Uist to Lewis.   1, 2, 8, 12, 15, 16

**C. ovalis** Good.
Oval Sedge.
Seisg Ughach.
Native.
Damp grassy places.
Barra to Lewis.   1–7, 9, 10, 12, 14–22, 27

**C. echinata** Murray
Star Sedge.
Seisg Reultach.
Native.
Damp moorland flushes, bogs and streamsides.
Berneray and Mingulay to Lewis and St Kilda.   2–7, 9–24, 27–29

**C. dioica** L.
Dioecious Sedge.
Seisg Aon-cheannach.
Native.
Flushes.
Barra to Lewis.   1, 3–5, 7, 9, 12, 16, 18, 19, 21

**C. pauciflora** Lightf.
Few-flowered Sedge.
Seisg nan Lusan Gann.
Native.
Acid bogs.
South Uist to Lewis but rare.   2, 3, 5, 7, 9, 12, 19

**C. pulicaris** L.
Flea Sedge.
Seisg na Deargainn.**
Native.
Bogs, streamsides and damp grassland.
Berneray and Mingulay to Lewis and St Kilda.   1–7, 9, 12–24, 27–29

**C. curta** Good.
White Sedge.
Seisg Bhàn.
Native.
Bogs and acid fens.
North Uist to Lewis.   1, 2, 8, 12, 15, 16

**C. lasiocarpa** Ehrh.
Slender Sedge.
Seisg Choilleanta.
Native.
Known from a few sites from North Uist to Lewis. North Uist *M. S. Campbell* 1937 (CGE), *J. W. Campbell* 1937 (BM); Harris *M. S. Campbell* 1959 (Perring, 1961); Uig, Lewis (Wilmott, 1948) (as C. *filiformis* auct. non L.).   3, 9 or 12, 16

**C. acutiformis** Ehrh.
Lesser Pond-sedge.
Isle of Baleshare, North Uist *J. W. H. Harrison* 1942 (K) conf. *Chater*

**C. rostrata** Stokes
Bottle Sedge.
Seisg Uisge.
Native.
Oligotrophic marshes and loch margins.
Vatersay to Lewis.   1–7, 9, 12, 16–19, 21, 22

**C. flacca** Schreber
Glaucous Sedge.
Seisg Liath-ghorm.
Native.
Damp grassland, banks and machair.
Mingulay to Lewis and St Kilda to North Rona.   1–9, 11–25, 27, 29

**C. panicea** L.
Carnation Sedge.
Seisg a' Chruithneachd.
Native.
Moorland, flushes and machair.
Mingulay to Lewis and St Kilda.   1, 3–7, 9–24, 27–29

**C. distans** L.
Distant Sedge.
Seisg Fhada-mach.
Native.
Saltmarshes.
Mingulay to Lewis and Monach Is.   3, 8, 12, 15, 16, 19, 21–24, 27, 29

**C. binervis** Sm.
Green-ribbed Sedge.
Seisg Fhèith-ghuirm.
Native.
Moorland.
Mingulay to Lewis, St Kilda to Shiant Is. 3–7, 9–13, 16–19, 21–24, 27–29
Var. *sadleri* (E. F. Linton) E. F. Linton has been recorded from some sites, but this variety, which differs in having a narrower utricle, longer beak and less distinct lateral veins is no longer separated.

**C. extensa** Good.
Long-bracted Sedge.
Seisg Anainn.
Native.
Upper levels of saltmarshes.
Sandray to Butt of Lewis.   1, 3, 8, 16, 19, 23

**C. hostiana** DC.
Tawny Sedge.

Seisg Odhar.
Native.
Fens and Flushes.
Sandray to Lewis and St Kilda.   1–3, 5, 7, 9, 11, 12, 15, 16, 18, 19, 21–23, 28

**C. × appeliana** Zahn
(**C. hostiana × demissa** ).
Occasional, with the parents. Literature record from Berneray (Harris) in Harrison (1941a). Specimens in BM and K from Barra, Pabbay and Harris.   9, 12, 13, 15, 21

**C. lepidocarpa** Tausch
Long-stalked Yellow-sedge.
Seisg Bhuidhe Fhad-chuiseagach.
Native.
Base-rich flushes.
Scattered and rarer than the next species but recorded from Vatersay to Lewis.   3, 4, 8, 9, 16, 21, 22

**C. demissa** Hornem.
Common Yellow-sedge.
Seisg Bhuidhe Choitcheann.
Native.
Flushes, loch margins and wet tracks.
Mingulay, Sandray and Flodday to Lewis and St Kilda.   3, 4, 6, 7, 9–19, 21–24, 27, 28

**C. serotina** Mérat
subsp. **serotina**
Small-fruited Yellow-sedge.
Seisg Bhuidhe nam Measan Beaga.
Native.
Loch margins, salt marshes and damp machair.
Berneray and Mingulay to Lewis and St Kilda.   1, 3, 4, 7–11, 15, 16, 18, 19, 21, 22, 24, 28

subsp. **pulchella** (Lönnr.) van Ooststr.
*C. scandinavica* E. W. Davies
Recorded from a few sites from Barra to Lewis but intergrades with the last subspecies so much that it is probably not worth maintaining.

**C. pallescens** L.
Pale Sedge.
Seisg Phreasach.
Native
Recorded from South Harris *W. S. Duncan* 1891 (BM); Glen Trollamarig, North Harris *W. A. Clark* 1941 (K) also recorded from Barra and North Uist *Shoolbred* (1895); South Uist Harrison (1941a); Lewis and Great

Bernera, Harrison (1957).   3, 4, 9, 12, 16, 19, 21

**C. pilulifera** L.
Pill Sedge.
Seisg Lùbach.
Native.
Dry moorland and dry banks.
Barra to Lewis and St Kilda.   2, 3, 5, 7, 9, 12, 16, 18, 19, 21, 28

**C. limosa** L.
Bog-sedge.
Seisg na Mòna.
Native.
Bogs and edges of lochans.
Sandray to Lewis.   1–5, 9–12, 16, 19, 23

C. magellanica Lam. (*C. paupercula* Michx)
Seisg na Bogaich.
was recorded from Tarbert, South Harris *Shoolbred* 1894 but redetermined as the preceding species, see Bennett (1897).

**C. caryophyllea** Latourr.
Spring Sedge
Seisg an Earraich.
Native.
Field record from Ersary, Barra by *Pankhurst* and *Chater* 1983; also recorded by Harrison (1941a) from Vatersay to Pabbay.   21–23.

**C. nigra** (L.) Reichard
*C. goodenowii* Gay
Common Sedge.
Gainnisg.
Native.
Marshes, flushes, loch margins and moorland.
Throughout the Islands.   1–16, 18–24, 25, 27–29
Wilmott (1938) examined the plant found in North Harris and published by A. Bennett as **C. spiculosa** var.**hebridensis** and concluded that it is a new variety **hebridensis** of C. goodenowii. It has unusually long and narrow, pointed glumes. The vars. **chlorostachya** Rchb., with long pale utricles which exceed the glumes and give the spikes a greenish colour, and **juncea** (Fries) Hyl. (= C. juncella auct.angl.), which has short rhizomes and a tussocky habit, are also recorded.

**C. bigelowii** Torrey ex Schweinitz
Stiff Sedge.
Dùr-sheisg.

Native.
Rocks, ledges and grassland on mountains. High ground from Barra to Lewis and St Kilda.   3, 8, 9, 12, 16, 19, 21, 28

Records from the Outer Hebrides by *J. W. H. Harrison* (see Harrison, 1945) for the following species are stated by Jermy, Chater and David (1982) to be erroneous. C. capitata (South Uist), C. remota (South Uist), C. chordorhiza (South Harris), C. lachenalii (Harris), C. rupestris (South Uist), C. microglochin (Harris) and C. norvegica (North Uist), specimens at K.

Carex vaginata Tausch was recorded from Beinn Mhor-Hecla Massif, South Uist by *J. W. H. Harrison*. The lack of suitable habitat and its distribution on the mainland make this extremely unlikely. Similar remarks apply to C. rariflora (Wahlenb.) Sm.

Similarly C. muricata was recorded from Vatersay, and C. vesicaria from Howmore district of South Uist, and cannot be credited without reconfirmation.

GLUMIFLORAE
*GRAMINEAE*
[*POACEAE*]

**Festuca** L.
**F. pratensis** Hudson
Meadow Fescue.
Fèisd Lòin.
Probably introduced.
Roadsides, sown grassland.
Barra, South Uist and Lewis.   1, 3, 5, 11, 16, 19, 21

**F. arundinacea** Schreber
Tall Fescue.
Fèisd Ard.
Native.
Roadsides.
Vatersay to Lewis but rare.   3, 11, 12, 16, 18, 21, 22

**F. rubra** L.
Red Fescue.
Fèisd Ruadh.
Native.
Saltmarsh, maritime heath, machair, moorland and mountains.
Throughout the Islands.   1–7, 9–13, 15, 16–25, 27–29

The taxonomy of the subspecific ranks is confused but specimens exist which have been named as follows;
subsp. **rubra**
subsp. **arenaria** (Osbeck) Syme and probably
subsp. **juncea** (Hackel) Sóo
subsp. **megastachys** Gaud. has been identified from Stornoway by *Hubbard*.
Specimens labelled F. juncifolia from St. Kilda and Barra at E have been redetermined by *Wilmott* as F. ovina var. arenaria.

**F. ovina** L.
Sheep's-fescue.
Feur-chaorach.
Native.
Scarce on moorland.
Berneray to Lewis and St Kilda to North Rona.   2–7, 9, 12, 13, 15, 16, 18–24, 25, 28

**F. vivipara** (L.) Sm.
Viviparous Fescue.
Native.
Common on moorland.
Berneray and Mingulay to Lewis and Shiant Is.   1, 3–7, 9, 10, 12, 13, 16–19, 21–24, 27
Wilmott in Campbell (1945) attempted to divide this rather variable species into several subspecies including subspp. **faeroensis, hebridensis, jemtlandica, norvegica, uigensis** and **vaagensis** but these are not accepted in modern treatments of the genus.

**Festuca longifolia** Thuill. and **F. glauca** Lam. were recorded by *J. W. H. Harrison* et al. and probably represent glaucous forms of F. rubra, perhaps subsp. pruinosa (Hackel) Piper.

**Lolium** L.
**L. perenne** L.
Perennial Rye-grass.
Breòillean.
Native.
Roadsides, waste places and improved pasture.
Berneray and Mingulay to Lewis, Shiant Is. and St Kilda.   2, 3, 4, 6, 7, 9–12, 14–22, 24, 27–29

**L. multiflorum** Lam.
Italian Rye-grass.
Breòillean Eadailteach.
Introduction.
Benbecula and Brevig, Barra, Harrison (1941a); North Uist, field records, *J. W. Clark*;

Lewis, Copping (1977) from Stornoway Harbour.
Probably under-recorded.   5, 16, 18, 21

**Vulpia** C. C. Gmelin
**V. bromoides** (L.) S. F. Gray
Squirreltail Fescue.
Native.
Field record from North Harris estate, 1979; walltop, North Uist *Shoolbred* (1895); Loch Eynort, South Uist and Eriskay (Harrison, 1941a) and Uig, Lewis (Wilmott, 1948).   3, 9, 16, 19, 20

**Desmazeria** Dumort.
(*Catapodium* Link)
**D. rigida** (L.) Tutin
*C. rigidum* (L.) C. E. Hubbard
Fern-grass.
Feur-raineach.
Native.
South Harris, Luskentyre Bank (Gimingham et al., 1948); Barra, cliff top, *Conacher* (1980); Mingulay, wall top of ruined house, *Chater* 1983 (BM).   12, 21, 24

**D. marina** (L.) Druce
*C. marinum* (L.) C. E. Hubbard, *D. loliacea* Nyman, *C. loliaceum* Link
Sea Fern-grass.
Feur Gainmhich.
Native.
Sandy areas.
Vatersay to Lewis.   1, 3, 4, 11–16, 21, 22

**Poa** L.
**P. annua** L.
Annual Meadow-grass.
Tràthach Bliadhnail.
Native.
Waste places, tracks and roadsides, dunes and arable land.
Probably in all zones.   1–12, 14–25, 27–29

Poa alpina L. was recorded from Allt Volagir, South Uist by Harrison (1941a), but needs to be confirmed.

**P. nemoralis** L.
Wood Meadow-grass.
Tràthach Coille.
Native.
Rare, Wall of Hotel, Tarbert, Harris *Shoolbred* (1895), Castle Park, Stornoway *M. S. Campbell* 1938 (BM) and *Copping* (1976) and Allt Volagir, South Uist *Warburg* 1947 (BM).   5, 9, 19

**P. pratensis** L.
Smooth Meadow-grass.
Tràthach an Lòin.
Native.
Rock ledges and walls.
Berneray to Lewis and St Kilda to North
Rona.   1–7, 9, 10, 12, 14–16, 18–22, 24, 25,
28, 29

**P. subcaerulea** Sm.
Spreading Meadow-grass.
Tràthach Sgaoilte.
Native.
Machair, maritime heath, pasture and moor-
land.
Barra to Lewis and St Kilda to North Rona.
1–3, 5, 8, 9, 12, 15, 16, 18, 19, 21, 22, 25, 28

**P. trivialis** L.
Rough Meadow-grass.
Tràthach Garbh.
Native.
Waste places, *Iris* patches and scrub.
Berneray and Mingulay to Lewis and St
Kilda.   1–13, 15–19, 21–24, 28, 29

**P. palustris** L.
Swamp Meadow-grass.
Tràthach Lèana.
Introduction.
Ditches.
Although recorded from Lewis, Berneray
and South Uist (Harrison et al., 1939; Clark
and Harrison, 1940 and Harrison, 1941a)
1939–1941; Ensay, *Morton* 1958, there is only
one specimen; ditch by Uig School, Uig,
Lewis *Wilmott* 1946 (BM) det. *Trist.* 3, 14,
15, 19

**P. chaixii** Vill.
Broad-leaved Meadow-grass.
Tràthach na Duilleige Leathainn.
Introduction.
Parkland.
Castle Park, Stornoway, Lewis *Copping*
(1976).   5

**Puccinellia** Parl.
**P. maritima** (Hudson) Parl.
Common Saltmarsh-grass.
Feur Rèisg Ghoirt.
Native.
Saltmarsh.
Barra to Lewis and Flannan Isle, St Kilda to
North Rona.   1–5, 7–10, 12, 15–21, 24, 25,
26, 29

**P. distans** (Jacq.) Parl.
subsp. **distans**
Rare. One record only, side of river at
bridge, Howmore, South Uist, *Pankhurst*
1983 det. *Melderis*.   19
subsp. **borealis** (Holmberg) W. E. Hughes
*P. capillaris* (Liljeblad) Jansen
Northern Saltmarsh-grass.
Native.
Saltmarsh.
Uig, Lewis *Crabbe* 1946 (BM), North Rona
*Atkinson* 1946 (BM) and *Gilbert* and *Holligan*
1976 (K), Stornoway Harbour and Butt of
Lewis *Copping* (1976).   1, 3, 25

**P. distans** subsp. **borealis** × **maritima**
*Gilbert* and *Holligan* also collected sterile
material which is the hybrid between the last
two taxa.
North Rona, 1976, 1979 (K) (see Gilbert and
Holligan, 1979)

**Dactylis** L.
**D. glomerata** L.
Cock's-foot.
Garbh-fheur.**
Native.
Roadsides and improved pastures.
Forms with viviparous spikelets occur.
Barra to Lewis.   2–5, 7, 9, 12, 15–22

**Cynosurus** L.
**C. cristatus** L.
Crested Dog's-tail.
Coin-fheur.
Native.
Machair, pastures, roadsides and dry moor-
land.
Berneray and Mingulay to Lewis and Shiant
Is.   1–7, 9–12, 14–22, 24, 27, 29

**Catabrosa** Beauv.
**C. aquatica** (L.) Beauv.
Whorl-grass.
Feur-sùghmhor.*
Native.
Wet ditches, sandy streams by sea.
Subsp. **minor** (Bab.) Perring & P.D.Sell is
recorded (as *C. aquatica* var. *littoralis*
Parnell) from the lea shore of Ben Lee by
*Shoolbred*, 1894 and from the coast near
Tigharry, both in North Uist, and from
Berneray in the Sound of Harris (all at BM).
Var. **grandiflora** Hack., with lemmas up to 4
mm, is recorded from Loch Paible, North
Uist, *J. W. Clark* 1975 (K) det. *Hubbard*.
Barra to Lewis. 1, 3, 4, 12–16, 18, 19, 21, 22, 29

**Briza** L.
**B. media** L.
Quaking-grass.
Conan Cumanta, Crith-fheur.
Doubtfully native.
Roadsides.
Rare, Northton, Harris 1948 *J. W. H. Harrison* (K), confirmed by field record 1983; E. foot of Ben Eoligarry, Barra *Davis* 1958 (E); Coll, Lewis *Clark* and *Copping* (1976). 1, 12, 21

**Glyceria** R. Br.
**G. fluitans** (L.) R. Br.
Floating Sweet-grass.
Mìlsean Uisge.
Native.
Ditches and loch margins.
South Uist to Lewis. 1–4, 6, 7, 9–19, 21, 22

**G. declinata** Bréb.
Small Sweet-grass.
Native.
Ditches and loch margins.
All field records. South Harris, Renish Point, *Dony* 1959, Rodel *Pankhurst* 1979; Parc, Loch Mor an Iaruinn, *Pankhurst* and *Chater* 1980; Eye peninsula, *Roe* 1959; Lewis, Five Penny Borve and Aird Dell, 1959. 1, 6, 7, 12

**Bromus** L.
**B. hordeaceus** L.
subsp. **hordeaceus**
*B. mollis* L.
Soft-brome.
Bròmas Bog.**
Native.
Arable fields, roadsides.
Berneray and Mingulay to Lewis. 3, 4, 6, 7, 9–12, 14–16, 18–22, 24, 27, 29
subsp. **thominii** (Hard.) Maire & Weiller
*B. thominii* Hard.
Lesser Soft-brome.
Native.
There appear to be a few genuine records of this taxon; Cliv Bay, Uig, Lewis *M. S. Campbell* 1939 (BM); Valain and Eoligarry, Barra *Davis* 1958 (E). 3, 21

**B. lepidus** Holmberg.
Slender Soft-brome.
Introduced.
Improved meadows.
South Uist, Garrynamonie and Smerclate, *Cadbury* 1978. 19

**Brachypodium** Beauv.
**B. sylvaticum** (Hudson) Beauv.

False Brome.
Bròmas Brèige.**
Native.
Stream ravines, sandy cliffs and plantations.
Mingulay to Lewis. 3, 4, 12, 19, 21–24

**Leymus** Hochst.
*Elymus* auct., non L.
**L. arenarius** (L.) Hochst.
Lyme-grass.
Taithean.
Native.
Foredunes.
Barra to Harris. 9, 13, 15, 16, 18, 19, 21

**Elymus** L.
*Agropyron* auct., non Gaertner
**E. repens** (L.) Gould
Common Couch.
Feur a' Phuint.
Native.
Dunes, shingle beaches and waste places.
Mingulay to Lewis and St Kilda. 1, 2, 4, 6, 7, 11, 12, 15–19, 21, 22, 24, 28

**E. farctus** (Viv.) Runemark ex Melderis
subsp. **boreali-atlanticus** (Simonet & Guinochet) Melderis
*A. junceiforme* (A. & D. Löve) A. & D. Löve
Sand Couch.
Glas-fheur.
Native.
Dunes.
Mingulay to Lewis and Monach Is. 1–4, 9, 11–13, 15, 16, 18, 19, 21–24, 29

**E. × littoreus** (Schumacher) Lambiñon
*Triticum littoreum* Schumacher
*T. laxum* Fries
*Elytrigia × litorea* (Schumacher) Hylander
*A. × laxum* (Fries) Tutin
*Elymus × laxum* (Fries) Melderis & McClintock
*E. farctus × repens*
Native.
Shingle and dunes.
Barra, Eoligarry jetty, *Pankhurst*, 1983 (BM); North Uist, Vallay Strand, *Williams* 1938 (BM), Hougharry, *J. W. Clark* 1973–5 (E); Lewis *M.S.Campbell* 1959 (Perring, 1961), Coll *Copping* (1976), L. Ordais, *McKean* 1985, BM and E. 1, 2, 9, 16, 21

**Hordeum** L.
**H. secalinum** Schreber
Meadow Barley.
Probably introduced.
Foot of steps to Coddie's House, Northbay, Barra *Davis* 1958 (E)

**Avena** L.
**A. fatua** L.
Wild-oat.
Coirce Fiadhain.
Introduction.
Apparently descreasing.
Lewis, Uig, *Wilmott* 1946 (BM) det. *Trist*;
South Harris, Lingara Bay, *Copping* 1975.
Harrison (1941a) also gives records for Great
Bernera, Berneray (Harris), and Eriskay.   3,
4, 12, 15, 20

**A. strigosa** Schreber
Bristle Oat, Bear.
Coirce Dubh.
Introduction.
Arable fields (as crop and as weed) and
waste places.
Barra to Lewis and St Kilda.   1, 9, 12, 14–16,
18, 19, 21, 28, 29

**Avenula** (Dumort.) Dumort.
(*Helictotrichon* Besser)
**A. pubescens** (Hudson) Dumort.
(*H. pubescens* (Hudson) Pilger)
Downy Oat-grass.
Feur Coirce.**
Native.
Machair.
Berneray and Mingulay to Lewis.   3, 9, 11,
12, 16, 18, 19, 21–24

**Arrhenatherum** Beauv.
**A. elatius** (L.) Beauv. ex J. & C. Presl
False Oat-grass.
Feur Coirce Brèige.**
Native.
Waste places,roadsides and ditchsides.
Mingulay to Lewis and St Kilda.   1, 3, 5–7,
10–12, 15–22, 24, 28

**Koeleria** Pers.
**K. macrantha** (Ledeb.) Schultes
*K. cristata* (L.) Pers., pro parte, *K. gracilis*
Pers.
Crested Hair-grass.
Cuiseag Dhosach.
Native.
Machair, maritime heath, dry banks and
moorland.
Berneray and Mingulay to Lewis and St
Kilda.   1–4, 9, 11, 12, 15, 16, 18–24, 28, 29
It is possible that **K. glauca** (Schrader) DC.
occurs in the Outer Hebrides but no speci-
mens have been seen. It can be distiguished
from the last by the non-flowering stalks
being convolute and densely silvery-scabrid

above instead of being flat or folded and
more or less smooth above.

Trisetum   flavescens   (L.)   Beauv.   was
recorded from St Kilda Macgillivray (1842)
but there is no specimen at K and the record
is unlikely.

**Deschampsia** Beauv.
**D. cespitosa** (L.) Beauv.
subsp. **cespitosa**
Tufted Hair-grass.
Cuiseag Airgid.
Native.
Mountain ledges and stream ravines.
For uniformity, this account uses the treat-
ment given in Clapham, Tutin and Moore.
However, a different modern classification is
given in Rich and Rich (1988).
Barra to Lewis.   1–3, 5–7, 9, 11, 12, 16, 18,
19, 21, 22
subsp. **alpina** (L.) Tzvelev
*D. alpina* (L.) Roemer & Schultes
Alpine Hair-grass.
Native.
High moorland and mountain ledges.
South Uist and Harris.   9, 19

**D. flexuosa** (L.) Trin.
Wavy Hair-grass.
Mòin-fheur.
Native.
Moorland, ravines and rock ledges.
Mingulay to Lewis and St Kilda.   1, 3–7, 9,
10, 12, 14–24, 27, 28

**D. setacea** (Hudson) Hackel
Bog Hair-grass.
Native.
Wet Bogs and loch margins.
South Uist to Lewis.   1, 3–5, 9, 12, 13, 16,
18, 19

**Aira** L.
**A. praecox** L.
Early Hair-grass.
Cuiseag an Earraich.
Native.
Dry rocky knolls.
Barra to Lewis and St Kilda to North Rona.
1–7, 9–12, 14–22, 24, 25, 27–29

**A. caryophyllea** L.
Silver Hair-grass.
Sìdh-fheur.*
Native.
Roadside banks.

Barra to Lewis.  1, 3, 4, 7, 9, 12, 15, 16, 18–21

**Hierochloë** R. Br.
**H. odorata** (L.) Beauv.
Holy-grass.
Feur Moire.*
Native.
Very rare. South Uist, field record at Askernish, Pitkin et al. (1983); Nunton, Benbecula *M. S. Campbell* 1936 (BM), and not seen there since.  18, 19
This is a protected species, listed in the Red Data Book, Perring and Farrell (1983).

**Anthoxanthum** L.
**A. odoratum** L.
Sweet Vernal-grass.
Borrach.
Native.
Pasture, moorland and maritime heath.
Berneray and Mingulay to Lewis and Shiant Is. to St Kilda.  1–7, 9–24, 27–29

**Holcus** L.
**H. lanatus** L.
Yorkshire Fog.
Feur a' Chinn Bhàin.**
Native.
Machairs, maritime heaths, pastures, moorland and mountains.
Berneray and Mingulay to Lewis and Shiant Is., St Kilda to North Rona.  1–24, 25, 27, 29

**H. mollis** L.
Creeping Soft-grass.
Mìn-fheur.
Native.
Ravines, scrub, cliffs.
Rodel, South Harris *Balfour and Babington* 1941; Benbecula and North Uist Shoolbred (1895) and additionally South Uist, Eriskay and Barra Harrison (1941a). Field records from Stornoway, Eishken, Barra and Grimsay. 5, 7, 12, 16–21

**Agrostis** L.
**A. canina** L.
Velvet Bent.
Fioran Mìn.
Native.
Acid grassland.
Common. Mingulay to Lewis and St Kilda to Shiant Is.  1–4, 6–24, 27–29

**A. vinealis** Schreber
*A. canina* subsp. *montana* (Hartman) Hartman

Brown Bent.
Fioran Badanach.
Native.
Mountain grassland.
Under-recorded due to confusion with the last species but does occur as in North Uist; Clachan, *Hunt* 1961 (K) and Hougharry, *J. W. Clark* 1975 (K).  16.

**A. vinealis × capillaris**
Occurs on Scalpay *Wilmott, M. S. Campbell* and *Bangerter* 1939 (BM) det. *Philipson* and Castle Park, Lewis *Copping* (1976) det. *Hubbard*

**A. capillaris** L.
*A. tenuis* Sibth., *A. vulgaris* With.
Common Bent.
Freothainn.
Native.
Acid grassland.
Barra to Lewis and St Kilda to North Rona. 1–7, 9, 11, 12, 15–24, 25, 27–29

**A. gigantea** Roth
Black Bent.
Probably introduced.
Roadsides.
North Uist, Baleshare, *Hunt* 1961 (K); Lewis, Acres Hotel, Stornoway, *Copping* (1976) and Loch Fasgro, *Noltie* (1987).  1, 2, 16

**A. stolonifera** L.
Creeping Bent.
Fioran.
Native.
Dune slacks, saltmarshes, loch and streamsides and waste places.
The var. **palustris** (Huds.) Farw., from wet places and floating in water, is recorded. Mingulay to Lewis and St Kilda to North Rona.  1–10, 12, 15–24, 25, 28

**Ammophila** Host
**A. arenaria** (L.) Link
Marram.
Muran.
Native.
Mobile dunes and sandy banks.
Mingulay to Lewis and Monach Is.  1–4, 9, 11–16, 18–24, 29

Calamagrostis epigeios (L.) Roth was recorded from South Uist, Harrison et al. (1942b), but no specimen has been seen.

**Phleum** L.
**P. pratense** L.

Timothy.
Feur Cait.
Native.
Roadsides and waste places.
Some records refer to subsp. **bertolonii**
(DC.) Bornm. distinguished from subsp.
**pratense** mostly by its smaller size.
Barra to Lewis 3, 4, 7, 9, 10, 12, 14, 16–19, 21,
22

**Alopecurus** L.
**A. pratensis** L.
Meadow Foxtail.
Fiteag an Lòin.
Native.
Fairly rare; North Uist, Ensay, Harris and
Lewis and Monach Isles.   2, 5, 7, 12, 14, 16,
29

**A. geniculatus** L.
Marsh Foxtail.
Fiteag Cham.
Native.
Ditches, wet waste places and pool margins.
Barra to Lewis and St Kilda.   1–7, 9–12,
14–22, 28, 29

A. alpinus Sm. has been recorded Hecla,
Beinn Mhor massif, South Uist *J. W. H.
Harrison* 1939, 1941 but has never been re-
found. There is a specimen at K.

**Phalaris** L.
**P. arundinacea** L.
Reed Canary-grass.
Cuiseagrach.
The garden form var. **picta** has been
recorded as an escape. In spite of its abun-
dance in some localities, it is apparently not
grown as a crop (Copping, 1978).
North Uist to Lewis.   1, 12, 16

**Phragmites** Adanson
**P. australis** (Cav.) Trin. ex Steudel
Common Reed.
Seasgan, Cuilc.
Native.
Marshes and loch margins.

Berneray and Mingulay to Lewis and the
Shiant Is.   1–5, 10, 12, 13, 16–24, 27

**Danthonia** DC.
(inc. *Sieglingia* Bernh.)
**D. decumbens** (L.) DC.
(*S. decumbens* (L.) Bernh.)
Heath-grass.
Feur Monaidh.
Native.
Moorland, grassland and flushes.
Berneray and Mingulay to Lewis and St
Kilda.   1–7, 9–19, 21–24, 27, 28

**Molinia** Schrank
**M. caerulea** (L.) Moench
Purple Moor-grass.
Fianach.
Native.
Moorland, flushes, bogs and maritime heaths.
Subsp. **altissima** (Link) Domin. (subsp.
**arundinacea** (Schrank) H. Paul) is a larger
plant with lemmas 3.5–5.5 mm (instead of
3–4) and leaf width 4–8.5 mm (instead of
1–7), and might be present in the Outer
Hebrides.
Berneray and Mingulay to Lewis and St
Kilda.   1–7, 9–24, 27–29

**Nardus** L.
**N. stricta** L.
Mat-grass.
Beitean, Riasg.
Native.
Moorland and mountains.
Berneray and Mingulay to Lewis, St Kilda to
North Rona.   3–7, 9, 10, 12, 15–24, 25, 27, 28

**Spartina** Schreber
**S. anglica** C. E. Hubbard
Common Cord-grass.
Native.
The specimen from Traigh Luskentyre,
South Harris *Muirhead* 1966 (E) labelled **S.** ×
**townsendii** H. and J. Groves is probably
this. Also reported from Seilebost by Law
and Gilbert (1986), from what may be the
same locality.   12

# Appendix

*List of herbaria searched for records for the Flora*

The collections of the following institutions were completely searched for material from the Outer Hebrides.

| Abbreviation | Name of institute |
|---|---|
| BM | British Museum (Natural History), London |
| CGE | University of Cambridge, Botany School |
| E | Royal Botanic Garden, Edinburgh |
| GL | University of Glasgow, Glasgow |
| K | Royal Botanic Gardens, Kew |
| OXF | University of Oxford, Botany School |

Some specimens are quoted from;

| | |
|---|---|
| BON | Museum and Art Gallery, Bolton |
| LINN | Linnean Society, London |
| LIV | Merseyside County Museum, Liverpool |
| LTR | University of Leicester, Leicester |
| MANCH | Manchester Museum, Manchester |
| MNE | Maidstone Museum and Art Gallery |
| NMW | Cardiff, National Museum of Wales |
| SUN | Sunderland Central Museum |

but no complete search was made of these collections.

Specimens collected by Mrs. *C. W. Murray* in N.Uist have been deposited at the Department of Botany, University of Aberdeen (ABD). Similarly, specimens collected by Mrs. *J.W.Clark* have been deposited at E.

## List of collector's names

J. Adams
J.R. Akeroyd
I.S. Angus
G.M. Ash
R. Atkinson
C.D. Austen
C.C. Babington
J. Balfour
J.H. Balfour
E.B. Bangerter
R.M. Barrington
M. Barron
R.H. Begg
A. Bennett
J. Bevan
K.B. Blackburn
H.J.M. Bowen
D. Bowman
J.T.B. Bowman
M. Braithwaite
E.N. Brockie
B.S. Brookes
D.C. Burlingham
F. Burnier
W.R. Bush
D.J. Cadbury
E. Cameron
M.S. Campbell
J.W. Campbell
J.F.M. Cannon

M.J. Cannon
D.F. Chamberlain
A.O. Chater
A.S. Cheke
P.M. Chorley
A.R. Clapham
J.W. Clark
W.A. Clark
A. Cleave
E.R.T. Conacher
R.B. Cooke
T.A. Cope
A. Copping
H.V. Corley
M. Coulson
J.A. Crabbe
G. Crompton
L. Cunningham
P. Cunningham
A. Currie
F.F. Darling
P.H. Davis
J.E. Dony
F. Druce
G.C. Druce
Miss U.K. Duncan
W.S. Duncan
T.T. Elkington
H. Fergusson
R.S.C. Ferreira

B.W. Fox
C.R. Fraser-Jenkins
H. Gibson
O.L. Gilbert
J. Gladstone
J. Goodwin
M. Goodwin
S. Gordon
G.G. Graham
H. Heslop Harrison
J. Heslop Harrison
J.W. Heslop Harrison
R.L. Hauke
C.C. Haworth
P.M. Holligan
J. Holub
F. Horsman
C.E. Hubbard
P.F. Hunt
B. Jonsell
E.F. Linton
J.D. Lovis
W. Macgillivray
J. Macrae
C.V.B. Marquand
J. Martin
M. Martin
H. McAllister
F. McDonald
D.R. McKean
R. Meinertzhagen
A. Melderis
W.F. Miller
J.K. Morton
C.W. Muirhead
J.M. Mullin
C.W. Murray
B. Nevison
J.A. Nicholson
H.J. Noltie
E. Norman
J. Ounstead

C.N. Page
R.J. Pankhurst
K.J.F. Park
A.M. Paul
F.H. Perring
C.P. Petch
N. Polunin
M.E.D. Poore
B.A. Poulter
C.D. Preston
H.W. Pugsley
R.E. Randall
D.A. Ratcliffe
J.E. Raven
B.W. Ribbons
A.J. Richards
J. Robarts
J.H. Roberts
N.K.B. Robson
R.G.B. Roe
P.D. Sell
W.A. Shoolbred
A.J. Silverside
J. Sinclair
A. Somerville
M. Stewart
N.F. Stewart
A.L. Still
B.T. Styles
G. Taylor
P. Taylor
D.M. Turner-Etlinger
U.A. Vincent
S.M. Walters
E.F. Warburg
M.M. Webster
C. West
I.A. Williams
A.J. Wilmott
P.F. Yeo
R. Young

## *The map zones for the Flora of the Outer Hebrides*

*Zone*

1. Northern N.Lewis

The Stornoway to Barvas road, continued by the track which leads from Barvas to the sea, is zone 1 and 2 boundary.

2. Southern N.Lewis

Zone 2 is separated from zone 5 by the road from Stornoway to Garynahine.

3. Uig

Boundary follows the parish boundary in the east and south, as used by the Flora of Uig.

4. Great Bernera, Little Bernera

5. SE Lewis

This bounded on the west by the Uig parish boundary, in the south by the Vigadale River, by the Stornoway-Tarbert road between Arivruaich and the head of Loch Erisort, and in the north by the road from Stornoway to Garynahine.

6. Eye peninsula
    The boundary is taken at the narrowest part of the causeway.
7. Park
    See remarks on zone 5.
8. Scarp
9. N.Harris
    Bounded in the north by the Uig parish boundary and the Vigadale River.
10. Scalpay and Scotasay
11. Taransay
12. South Harris
13. Pabbay and Shillay
14. Sound of Harris
15. Berneray and Borneray

16. North Uist
17. Grimsay and Ronay
18. Benbecula
19. South Uist
20. Eriskay
21. Barra
22. Vatersay and Muldoanich
23. Sandray and Pabbay
24. Mingulay and Berneray
25. North Rona and Sula Sgeir
26. Flannan Is.
27. Shiant Is.
28. St.Kilda
29. Monach Is. and Heisker

# Index of place names mentioned in the Flora

NB. Different maps may have different spellings for the same place names. Some of the variations are included below.

| Locality | Zone number | Nat. Grid reference | Locality | Zone number | Nat. Grid reference |
|---|---|---|---|---|---|
| Cheese Bay | 16 | NF9673 | Gravir | 7 | NB3815 |
| Clachan | 16 | NF8163 | Great Bernera (island) | 4 | NB13 |
| Clettraval | 16 | NF7471 | Greian Head | 21 | NF6404 |
| Clisham | 9 | NB1507 | Gress River | 1 | NB44 |
| Cliv Bay | 3 | NB0836 | Grimersta | 3 | NB2130 |
| Cnoc an Fhithich | 21 | NF6504 | Grimsay (island) | 17 | NF85 |
| Coll | 1 | NB4640 | Grogarry | 19 | NF7739 |
| Coppay (island) | 12 | NF99 | Grosebay | 12 | NG1592 |
| Corodale Bay, Fuday | 21 | NF7307 | | | |
| Corrie Dubh | 19 | NF7932 | Halaman Bay | 21 | NF6400 |
| Cracaval | 3 | NB0225 | Hartavagh | 19 | NF8315 |
| Cravadale | 9 | NB0212 | Hartaval | 21 | NF6800 |
| Creagan Leathan | 9 | NB0713 | Haskeir (island) | 29 | NF68 |
| Creagorry | 18 | NF7948 | Heaval | 21 | NL6799 |
| Crossdougal | 19 | NF7520 | Hecla | 19 | NF8234 |
| Crowlista | 3 | NB0433 | Heisinish | 20 | NF8009 |
| Cuier | 21 | NF6703 | Heisker (Monach Is.) | 29 | NF66 |
| | | | Hermetray | 16 | NF97 |
| Dalbeg | 2 | NB2245 | Hoe Beg | 16 | NF9574 |
| Daleburgh = Daliburgh | 19 | NF7521 | Horgabost | 12 | NG0596 |
| Daliburgh | 19 | NF7521 | Hougharry | 16 | NF7071 |
| Deasker (island) | 29 | NF6466 | Howmore | 19 | NF7636 |
| Dremisdale | 19 | NF7537 | Hushinish | 9 | NA9812 |
| Drinishader | 12 | NG1794 | | | |
| | | | Kildonan | 19 | NF7427 |
| Ersary | 21 | NL7099 | Killegray (island) | 14 | NF98 |
| Eachkamish | 16 | NF7959 | Kilpheder | 19 | NF7419 |
| Eilean a' Mhorain | 16 | NF8379 | Kirkibost (island) | 16 | NF7865 |
| Eilean Molach | 3 | NA9932 | Kirivick | 2 | NB1941 |
| Eishken | 7 | NB3212 | Kneep | 3 | NB0936 |
| Enaclete | 3 | NB1228 | Knockintorran | 16 | NF7367 |
| Ensay (island) | 14 | NF9786 | Kyles Paible | 16 | NF7466 |
| Eoligarry | 21 | NF7007 | Kyles Scalpay | 9 | NG2198 |
| Eoropie | 1 | NB6154 | Kyles Stuley | 19 | NF8123 |
| Eriskay (island) | 20 | NF71 | | | |
| | | | Laxadale Burn | 9 | NB1800 |
| Five Penny Borve | 1 | NB4056 | Lee Hills = Ben Lee | 16 | NF9164 to 9366 |
| Fuday (island) | 21 | NF70 | Leverburgh | 12 | NG0186 |
| | | | Lingadale | 9 | NB1601 |
| Galson | 1 | NB4458 | Lingara Bay | 12 | NG0684 |
| Garrynahine | 2/5 | NB2331 | Linique | 19 | NF7546 |
| Garrynamonie | 19 | NF7416 | Little Bernera (island) | 4 | NB13 |
| Gasker | 8 | NA8711 | Little Loch Borve | 15 | NF9181 |
| Geiraha | 1 | NB5349 | Little Loch Roag | 3 | NB1227 |
| Geoan Dubh | 9 | NB1402 | Loch a' Chama | 3 | NB0324 |
| Gerinish | 19 | NF7841 | Loch a' Chlachain | 19 | NF7531 |
| Geshader | 3 | NB1131 | Loch a' Mhorghain | 9 | NB1504 |
| Gillaval Glas | 9 | NB1502 | Loch an Armuinn | 16 | NF9074 |
| Gill Laxdale | 12 | NG1096 | Loch an Duin | 19 | NF7429,7415 |
| Gill Meodale | 12 | NG0595 | Loch an Eilean | 19 | NF7416,7423, |
| Gleann Shranndabhal | 12 | NG0285 | | | 7537 |
| Glen Beasdale = | | | Loch an Rubha Ardvule | 19 | NF7129 |
| Beesdale | 12 | NB1100 | Loch an Tairbh Duinn | 12 | NG1399 |
| Glen Laxadale | 9 | NB1803 | Loch na Berie | 3 | NB1035 |
| Glen Meavaig | 9 | NB0908 | Loch na Clacha-mora | 19 | NF7521 |
| Glen Skeaudale | 9 | NB1403 | Loch na Doirlinn | 21 | NF6400 |
| Glen Tealasdale | 3 | NB0022 | Loch na Liana Moire | 18 | NF7653 |
| Glen Trollamarig | 9 | NB2001 | Loch na Liana Moire | 19 | NF7320 |
| Glen Village | 21 | NL6798 | Loch na Tanga | 19 | NF7323 |
| Goulaby Burn | 16 | NF8876 | Loch nam Buard | 29 | NF6361 |

| Locality | Zone number | Nat. Grid reference | Locality | Zone number | Nat. Grid reference |
|---|---|---|---|---|---|
| Loch Aird an Sgairbh | 19 | NF7326 | Lochboisdale | 19 | NF7919 |
| Loch Altabrug | 19 | NF7434 | Lochmaddy | 16 | NF9168 |
| Loch Ardvule | 19 | NF7129 | Lochskipport | 19 | NF8238 |
| Loch Arnol | 2 | NB3048 | Luskentyre | 12 | NG0799 |
| Loch Bee | 19 | NF74 | Luskentyre Banks | 12 | NG0699 |
| Loch Bhruist | 15 | NF9182 | | | |
| Loch Boisdale | 19 | NF81 | Maaruig River | 9 | NB1705 |
| Loch Bru | 16 | NF8973 | Machair Robach | 16 | NF8676 |
| Loch Ceann a' Bhaigh | 19 | NF7630 | Mangersta | 3 | NB0031 |
| Loch Cistavat | 12 | NG0395 | Manish | 12 | NG1089 |
| Loch Corodale | 19 | NF8333 | Marrival | 16 | NF8070 |
| Loch Drinishader | 12 | NG1695 | Mealasta | 3 | NA9924 |
| Loch Druidibeg | 19 | NF73 | Mealasta (island) | 3 | NA92 |
| Loch Eilaster | 2 | NB2238 | Mealisval | 3 | NB0226 |
| Loch Erisort | 5/7 | NB32 | Meavaig | 9 | NB0906 |
| Loch Eynort | 19 | NF8026 | Melbost Farm | 1 | NB4632 |
| Loch Fada | 16 | NF8770 | Miavaig | 3 | NB0834 |
| Loch Fasgro | 2 | NB2141 | Mingulay (island) | 24 | NL58 |
| Loch Grogarry | 16 | NF7171 | Moor Hill | 14 | NF9283 |
| Loch Hallan | 19 | NF7322 | Muldoanich (island) | 22 | NL69 |
| Loch Kearsinish | 19 | NF7916 | Mullach an Langa | 9 | NB1409 |
| Loch Kildonan | 19 | NF7328 | Mullach an Reidhachd | 9 | NB090143 |
| Loch Langavat | 3/5 | NB11 | | | |
| Loch Laxavat Ard | 2 | NB2438 | Na-Baighe-Dubha | 19 | NF7729 |
| Loch Laxdale | 12 | NG1096 | Ness | 1 | NB56 |
| Loch Leodasay | 16 | NF8063 | Newton | 16 | NF8977 |
| Loch Leurbost | 5 | NB3824 | Newton Ferry | 16 | NF8978 |
| Loch Maddy | 16 | NF96 | Nisabost | 12 | NG0597 |
| Loch Mhor, Baleshare | 16 | NF7962 | North Glendale | 19 | NF7917 |
| Loch Mor an Iaruinn | 7 | NB3019 | North Tolsta | 1 | NB5347 |
| Loch Obisary | 16 | NB86 | Northton | 12 | NF9989 |
| Loch Odhairn = | | | Nunton | 18 | NF7653 |
| L.Ouirn | 7 | NB4014 | | | |
| Loch Ollay | 19 | NF7631 | Oban Uaine | 18 | NF8447 |
| Loch Ordais | 2 | NB2848 | Obbe | 12 | NG0186 |
| Loch Paible | 16 | NF7268 | Oreval | 9 | NB0111 |
| Loch Portain | 16 | NF9471 | Ormaclett | 19 | NF7431 |
| Loch Roag | 3 | NB13 | Ormiclate = Ormaclett | 19 | NF7431 |
| Loch St. Clair | 21 | NL6499 | Oronsay (island) | 16 | NF8475 |
| Loch Sandary | 16 | NF7368 | | | |
| Loch Sandavat | 4 | NB1537 | Pabbay (island) | 13 | NF88 |
| Loch Scarie | 16 | NF7170 | Pabbay (island) | 23 | NL68 |
| Loch Seaforth | 7/9 | NB20 | Paible | 16 | NF7268 |
| Loch Shell | 7 | NB3110 | Peinylodden | 18 | NF7754 |
| Loch Skiport | 19 | NF8338 | Pennylodden = | | |
| Loch Snigisclett | 19 | NF8024 | Peinylodden | 18 | NF7754 |
| Loch Sniogravat | 29 | NF6360 | Pollachar | 19 | NF7414 |
| Loch Stiapavat | 1 | NB5264 | Port Sto | 1 | NB5265 |
| Loch Stilligary | 19 | NF7638 | Portvoller | 6 | NB5667 |
| Loch Stromore (an | | | | | |
| Strumore) | 16 | NF8969 | Renish Point | 12 | NG0481 |
| Loch Teanga | 19 | NF8139 | Rodel | 12 | NG0483 |
| Loch Torcusay | 18 | NF7653 | Ronay (island) | 17 | NF85 |
| Loch Torornish | 19 | NF7330 | Roneval | 12 | NG0486 |
| Loch Trollamarig | 9 | NB2201 | Rubha Ardvule | 19 | NF7029 |
| Loch Uacraich | 18 | NF7751 | Rueval | 18 | NF8253 |
| Loch Uamavat = | | | Risgary | 15 | NF9381 |
| L.Uamadale? | 12 | NG1398 | | | |
| Loch Urrahag | 2 | NB3247 | Sandray (island) | 23 | NL69 |
| Lochan nam Faoileann | 21 | NF7001 | Scaladale | 9 | NB1709 |

| Locality | Zone number | Nat. Grid reference | Locality | Zone number | Nat. Grid reference |
|---|---|---|---|---|---|
| Scalpay (island) | 10 | NG29 | Teanamachar | 16 | NF7761 |
| Scarasta = Scarista | 12 | NG0093 | Tigharry | 16 | NF7171 |
| Scarastavore | 12 | NG0092 | Tirga Mor | 9 | NB0511 |
| Scarista | 12 | NG0093 | Toa Galson | 1 | NB4560 |
| Scarp (island) | 8 | NA91 | Tolsta | 1 | NB54 |
| Scotasay (island) | 10 | NG19 | Tolsta Head | 1 | NB5647 |
| Seaforth Island | 7 | NB21 | Tomnaval | 9 | NB1607 |
| Seilebost | 12 | NG0696 | Traigh an Taoibh | | |
| Sgaoth Aird | 9 | NB1604 | Thuath | 12 | NF99 |
| Sgurr Scaladale | 9 | NB1608 | Traigh Mhor | 1 | NB5448 |
| Shawbost | 2 | NB2546 | Traigh Luskentyre | 12 | NG0797 |
| Sheaval | 19 | NF7627 | Traigh Scurrival | 21 | NF7008 |
| Shiant Islands | 27 | NG94 | Trumisgarry | 16 | NF8674 |
| Shillay (island) | 13 | NF8791 | | | |
| Sidinish | 16 | NF8763 | Uachdar | 18 | NF7955 |
| Skeaudale River | 9 | NB1403 | Uamasclett | 12 | NG1399 |
| Skigersta | 1 | NB5461 | Udal | 16 | NF8278 |
| Smerclate | 19 | NF7415 | Uig | 3 | NB03 |
| Sollas | 16 | NF8074 | Uisgnaval Beg | 9 | NB1108 |
| Sponish | 16 | NF8864 or 9269 | Uisgnaval More | 9 | NB1208 |
| Sron Scourst | 9 | NB1009 | Ulladale | 9 | NB0814 |
| Stilligarry | 19 | NF7638 | Ullaval | 9 | NB0811 |
| Stockay (island) | 29 | NF6663 | | | |
| Stoneybridge | 19 | NF7433 | Valain | 21 | NF6905 |
| Stornoway | 5 | NB43 | Vallay (island) | 16 | NF7776 |
| Strone (=Sron) Scourst | 9 | NB1009 | Vallay Strand | 16 | NF7875 |
| Stuley islands | 19 | NF82 | Valtos | 3 | NB0936 |
| Suainaval | 3 | NB0730 | Valtos Glen | 3 | NB0734 |
| Sythe | 11 | NB0202 | Vaslain | 21 | NF6904 |
| | | | Vatersay (island) | 22 | NL69 |
| Taireval | 3 | NB0024 | Vatisker Point | 6 | NB4939 |
| Tangusdale | 21 | NF6400 | | | |
| Taransay (island) | 11 | NB00 | West Loch Tarbert | 9/12 | NB00 |
| Tarbert | 9/12 | NB1500 | Wiay (island) | 18 | NF84 |
| Tealasdale River | 3 | NB0022 | | | |

# BIBLIOGRAPHY of the flora of particular ISLANDS

The zone numbers are given in brackets.

BALESHARE, NORTH UIST (16)
  Clark (1939c)

BARRA (21)
  Burnett (1964)
  Campbell (1936)
  Conacher (1980)
  Dawson (1936)
  Forrest et al. (1936)
  Harrison & Harrison (1950a)
  King's College (1939)
  Mackenzie 1936
  MacLeod (1949)
  Scott (1894)
  Somerville (1889, 1890)
  Watson (1939a, b)

Watson & Barrow (1936)
Wilmott (1939)

BERNERAY, BARRA (24)
  Cheke and Reed (1987)
  Clark (1938)
  Clark (1939c)
  Smith (1987)

BERNERAY, SOUND OF HARRIS (15)
  Clark & Harrison (1940)

BORERAY (ST. KILDA) (28)
  Duncan, Bullock and Taylor, (1981)

CAMPAY (4)
  Currie (1981b)

CAUSAMUL (16)
  Atkinson (1954)

COPPAY (12)
Harrison (1954a)

DEASKER(29)
Atkinson (1954,1980)

ENSAY, SOUND OF HARRIS (14)
Harrison & Harrison (1950b)

ERISKAY(20)
Harrison et al (1939)

FLANNAN ISLES (26)
Bennett (1907)
Gilbert & Wathern (1976)

FUDAY, BARRA (21)
Harrison et al. (1939)

GASKER (8)
Atkinson & Roberts (1952)
Atkinson(1980)

GREAT BERNERA, UIG (4)
Clark & Harrison (1940)
Crummy (1982)
Harrison (1957)

GRIMSAY (17)
Harrison & Harrison (1950a)

HASKEIR (29)
Atkinson (1954,1980)

KILLEGRAY, SOUND OF HARRIS (14)
Harrison & Harrison (1950b)

LITTLE BERNERA (4)
Clark & Harrison (1940)
Crummy (1982)
Harrison (1957)

MINGULAY, BARRA(24)
Buxton (1987)
Cheke and Reed (1987)
Clark (1938)
Clark (1939c)

MONACH ISLES (29)
Clark (1939c)
Perring & Randall (1972)

NORTH RONA (25)
Atkinson (1940, 1946)
Barrington (1885)

Darling (1939)
Gilbert, Holligan & Holligan (1973)
Gilbert & Holligan (1979)
McVean (1961)

PABBAY (BARRA) (23)
Cheke and Reed (1987)

PABBAY, SOUND OF HARRIS(13)
Atkinson (1980)
Clark & Harrison (1940)
Elton (1938)

RONAY (17)
Harrison & Harrison (1950a)

ST. KILDA (HIRTA) (28)
Barrington (1886)
Boyd in Small (1979)
Brathay (1971)
Gladstone (1928)
Gwynne, Milner & Hornung (1974)
Macgillivray (1842)
McVean (1961)
Petch (1932, 1933a)
Poore & Robertson (1949)
Praeger (1897)
Turrill (1927)

SCALPAY, HARRIS (10)
Campbell (1944)

SHIANT ISLANDS (27)
Atkinson (1954)
Harrison (1951b)
Harrison (1953)

SHILLAY, SOUND OF HARRIS (13)
Harrison (1954a)
Atkinson (1980)

STULEY ISLANDS (19)
Harrison & Harrison (1950a)

SULA SGEIR (25)
Atkinson (1940)
McVean(1961)
Gilbert, Holligan & Holligan (1973)

TARANSAY (11)
Clark (1939c)
Harrison (1954a)

# Bibliography

Manuscripts are to be found in the library of the British Herbarium (reprint collection) at the BM, unless otherwise stated.

**Adam, P.** 1981. The vegetation of British saltmarshes. *New Phytologist* **88**: 143–196.

**Akeroyd, J. R. & Curtis, T. C. F.** 1980. Some observations on the occurrence of machair in western Ireland. *Bull. Irish. Biogeog. Soc.* **4**: 1–12.

**Allen, D. E.** 1951. Abstracts from literature. 102–104, 110, Inner and Outer Hebrides (review of J. H. Harrison, 1948). *Watsonia* 2: 129–131.

———— 1952. *Cakile edentula* (Bigel.) Hook. in Britain. *Watsonia* 2: 282–283.

———— 1954. Variation in *Peplis portula* L. *Watsonia* 3: 85–91.

**Anderson, M. L.** 1967. A history of Scottish forestry. Thomas Nelson & Sons, London & Edinburgh.

**Angus, S.** 1987. Lost woodlands of the Western Isles. *Heb. Nat.*, 9, 24–30.

**Annandale, C.** (Ed.) (1890–) 1893. The Popular Encyclopedia. New Issue Revised. **14**: 58. Blackie & Son Ltd.

**Atkinson, R.** 1940. Notes on the Botany of North Rona & Sula Sgeir. *Trans. & Procs. Bot. Soc. Edinb.* 33: 52–60 (with a letter re *Poa pratensis*).

———— 1946. List of species recorded on N. Rona for 28.7.1946. MS.

———— 1954. 1) List of plants from Deasker and Causamul. MS.

———— 2) List of plants from Shiant Islands. MS.

———— 1980. Shillay and the seals. Collins and Harvill Press, London. 167pp.

**Atkinson, R. & Roberts, B.** 1951. Notes on the islet of Gasker. *Scot. Nat.* **64**: 129–37.

**Babington, C. C.** 1897. Memorials, journal and botanical correspondence of Charles Cardale Babington. Cambridge, MacMillan & Bowes.

**Baker, E. G.** 1898. Some British Violets, II. *J. Bot.* **39**: 224.

**Balfour, J. H. & Babington, C. C.** 1842. Account of a Botanical Excursion to Skye and the Outer Hebrides, during the month of August 1841 . . . and remarks on the plants observed by them in the island of N. Uist, Harris and Lewis. *Ann. & Mag. Nat. Hist.* **8**: 541–542.

**Balfour, J. H., Babington C. C.** 1844. An account of the vegetation of the Outer Hebrides. *Trans. Bot. Soc. Edinb.* **1**: 133–154.

**Bangerter, E. B.** 1939. My Thoughts go Back (Outer Isles). MS.

**Barrington, R. M.** 1885. In 'Further notes on North Rona' by J. A. Harvie-Brown, *Proc. Roy. Phys. Soc. Edinb.* 9: 289.

———— 1886. Notes on the Flora of St. Kilda. *J. Bot.* 24: 213–216.

**Bateman, R. M. & Denholm, I.** 1989. A reappraisal of the British and Irish dactylorchids, 3. The spotted-orchids. *Watsonia* 17: 319–349.

**Beeby, W. H.** 1887. On the flora of Shetland. *Scot. Nat.* 9: 20–32.

———— 1888. On the flora of Shetland. *Scot. Nat.* 9: 211.

**Bennett, A.** 1886 (–1891). Additional Records of Scotch plants for 1886. *Scot. Nat.* 1886–1891.

———— 1889. Notes on the Flora of the Outer Hebrides. *Trans. Nat. Hist. Soc. Glasgow* (new series), 1889. 3: 37–41.

———— 1891. Records for Scottish Plants for 1890 additional to 'Topographical Botany' Ed. 2 (1883). *Ann. Scot. Nat. Hist.* 1891: 188–189.

———— 1892a. Contributions towards a flora of the Outer Hebrides, I and II. *Ann. Scot. Nat. Hist.* 1892: 56–64 and 240–243.

———— 1892b. Records for Scottish Plants for 1891 additional to 'Topographical Botany' Ed. 2 (1883). *Ann. Scot. Nat. Hist.* 1892: 126–127.

———— 1893. Records for Scottish Plants for 1892 additional to 'Topographical Botany' Ed. 2 (1883). *Ann. Scot. Nat. Hist.* 1893: 101.

———— 1893. *Juniperus intermedia* Schur. in Scotland. *Ann. Scot. Nat. Hist.* 1893: 250–251.

———— 1894a. Records for Scottish Plants for 1893 additional to 'Topographical Botany' Ed. 2 (1883). *Ann. Scot. Nat. Hist.* 1894: 163.

———— 1894b. *Chrysosplenium oppositifolium* in the Outer Hebrides. *Ann. Scot. Nat. Hist.* 1894: 186.

———— 1895. Records for Scottish Plants for 1894 additional to 'Topographical Botany' Ed. 2 (1883). *Ann. Scot. Nat. Hist.* 1895: 117–118.

———— 1896. *Elatine hexandra* DC. in the Outer Hebrides. *Ann. Scot. Nat. Hist.* 1896: 63–64.

———— 1897. *Carex magellanica* L. in the Outer Hebrides. *Ann. Scot. Nat. Hist.* 1897: 188–190.

———— 1905a. Contributions towards a flora of the Outer Hebrides, III. *Ann. Scot. Nat. Hist.* 1905: 164–174.

———— 1905b. Supplement to *Topographical Botany* Ed.2. *J. Botany*, **48**, supplement.

———— 1907. The plants of the Flannan Islands. *Ann. Scot. Nat. Hist.* 1997: 187.

———— 1910. Contributions towards a flora of the Outer Hebrides, IV. *Ann. Scot. Nat. Hist.* 1910: 165–170.

———— 1911. *Pyrola uniflora* L. and *Valeriana dioica* L. in the Outer Hebrides. *Ann. Scot. Nat. Hist.* 1911: 185–186.

**Bennett, A., Salmon, C. E. & Matthews, J. R.** 1929–1930. Second supplement to *Topographical Botany* Ed. 2. *J. Botany*, **67,68**, supplement.

**Bennett, K. D**. 1984. The post-glacial history of *Pinus sylvestris* in the British Isles. *Quaternary Science Reviews* 3: 133–155.

———— 1985. The spread of *Fagus grandifolia* across eastern North America. *Journal of Biogeography* **12**: 147–164.

**Bennett, K. D. & Fossitt, J.** 1988. A stand of birch by Loch Eynort, South Uist, Outer Hebrides. *Trans. Bot. Soc. Edinb.* **45**: 245–252.

**Beveridge, E.** 1911. North Uist, Archaeology & Topography (Botany pp.12–13) William Brown, Edinburgh pp. xxvi and 348.

———— 1926. The submerged forest and peat off Vallay, *North Uist. Scot. Nat.* **157**: 24–25.

**Biagi, J. A, Chamberlain, D. F, Hollands, R. C, King, R. A, & McKean, D. R.** 1985. Freshwater macrophyte survey of selected lochs in Lewis and Harris. Royal Botanic Garden, Edinburgh. Internal report for N.C.C.

**Bibby, J. S., Douglas, H. A., Thomasson, A. J. & Robertson, J. S.** 1982. *Land capability classification for agriculture.* Soil Survey of Scotland Technical Monograph. The Macaulay Institute for Soil Research, Aberdeen.

**Birks, H. J. B**. 1973. *Past and present vegetation of the Isle of Skye – a palaeoecological study.* Cambridge University Press, London, pp. xii, 415.

———— 1977. The Flandrian forest history of Scotland: a preliminary synthesis. In *British Quaternary Studies – Recent Advances* (Ed. **F. W. Shotton**): 119–135. Clarendon Press, Oxford.

———— 1988. Long-term ecological change in the British uplands. In *Ecological Change in the Uplands* (Eds. **M. B. Usher & D. B. A. Thompson**): 37–56. Blackwell Scientific Publications, Oxford.

———— 1987. The vegetational context of Mesolithic occupation on Oronsay and adjacent areas. In *Excavations on Oronsay – Prehistoric Human Ecology on a Small Island* (Ed. **P. Mellars**): 71–77. Edinburgh University Press, Edinburgh.

**Birks, H. J. B. & Madsen, B. J**. 1979. Flandrian vegetational history of Little Loch Roag, Isle of Lewis, Scotland. *J. Ecol.* **67**: 825–842.

**Birks, H. J. B. & Williams, W**. 1983. Late Quaternary vegetational history of the Inner Hebrides. *Proc. Roy. Soc. Edinb.* **83B**: 269–292.

**Birnie, J**. 1983. Tolsta Head: further investigations of the interstadial deposit. *Quaternary Newsletter* **41**: 18–25.

**Birse, E. L**. 1971. *Assessment of climatic conditions in Scotland. 3. The bioclimatic sub-regions.* The Macaulay Institute for Soil Research, Aberdeen.

**Birse, E. L. & Dry, F. T**. 1970. *Assessment of climatic conditions in Scotland, 1: based on accumulated temperatures and potential water deficit.* The Macaulay Institute for Soil Research, Aberdeen.

**Birse, E. L. & Robertson, L**. 1970. *Assessment of climatic conditions in Scotland, 2: based on exposure and accumulated frost.* The Macaulay Institute for Soil Research, Aberdeen.

**Birse, E. L. & Robertson, J. S**. 1976. *Plant communities and soils of the lowland and southern upland regions of Scotland.* The Macaulay Institute for Soil Research, Aberdeen.

**Blackburn, K. B**. 1940. Hebridean Pondweeds. *Vasculum* **26**: 25

———— 1946. On a peat from the Island of Barra, Outer Hebrides. *New Phytol.* **45**: 44–49.

Blake, J. L. 1966. Trees in a treeless island. *Scottish Forestry* **20**: 37–47.

Bohncke, S. J. P. 1988. Vegetation and habitation history of the Callanish area, Isle of Lewis, Scotland. In *The Cultural Landscape – past, present, and future* (Eds. **H. H. Birks, H. J. B. Birks, P. E. Kaland, & D. Moe**): 445–461. Cambridge University Press, Cambridge.

Bowman, J. T. B. & Bowman D. 1979. *Diphasiastrum alpinum* (L.) Holub in Harris. *Watsonia* **12**: 333.

Boyd, J. M. 1979a. Natural History, in *A St.Kilda handbook* (Ed. **A. Small**). National Trust for Scotland.

————— 1979b. The natural environment of the Outer Hebrides. *Proc. Roy. Soc. Edinb.* **77B**: 3–19.

Brathay Exploration Group. 1972. Field studies on St.Kilda, 1971. Field Studies report No. 20, Brathay Hall Trust, Ambleside, Westmorland.

Brown, A. 1986. Upland Survey Project: daily field reports on North Harris. Unpublished reports, Nature Conservancy Council.

B.S.B.I. 1951. Plant Records. *Watsonia* **2**: 58.

————— 1974. Plant Records. *Watsonia* **10**: 187.

————— 1970. Plant Records. *Watsonia* **8**: 53.

————— 1980. Plant Records. *Watsonia* **13**: 131.

Buchan, A. 1752. A description of Saint Kilda. Republished 1974, Stansfield, Fortrose.

Burnett, J. H. (Ed.) 1964. The Vegetation of Scotland. Oliver & Boyd, Edinburgh & London. Table 12: Synoptic Table showing Composition of Seral and relatively stable dune Communities (Harris, Barra, Lewis).

Buxton, N. E. 1987. Report on the vegetation of Mingulay. *Royal Air Force Ornithological Society J.* **17**: 64–77.

Caird, J. B. 1979. Land Use in the Uists since 1800. *Proc. Roy. Soc. Edinb.* B. **77**: 505–527.

Cameron, J. 1883. *Gaelic Names of Plants* (Scottish and Irish): Collected and Arranged in Scientific Order, with Notes of their Etymology, their Uses, Plant Superstitions etc., among the Celts, with Copious Gaelic, English, and Scientific Indices. Blackwood, Edinburgh & London.

Campbell, J. L. 1936. *The Book of Barra*: list of plants observed by Edinburgh University Biol. Soc. July 1935: 309–317. London, George Routledge & Sons Ltd.

Campbell, J. W. 1946. The food of the Widgeon and Brent Goose. *British Birds* **39**: 194–200, 226–232.

Campbell, M. S. 1936. Botanising in Benbecula. MS.

————— 1937a. Three weeks botanising in the Outer Hebrides. *B.E.C. 1936 Report* **11**: 304–318.

————— 1937b. Further Botanising in the Outer Hebrides *B.E.C. 1937 Report* **11**: 536–560.

————— 1939a. Botanical Expedition to the Outer Hebrides. July-August 1939. MS (for Stornoway Gazette).

————— 1939b. Notes on Outer Hebrides Records. *J. Bot.* **77**: 63.

————— 1940. Notes on Outer Hebridean Flora. *J. Bot.* **77**: 101–102.

————— 1944. A visit to Scalpay (Harris) V.C.110. *B.E.C Report* **12**: 543–545.

————— 1945. *The Flora of Uig (Lewis). A Botanical Exploration.* T. Buncle & Co. Ltd., Arbroath. pp 63.

————— 1947. Typed list of Rubus determinations due to W. C. R. Watson in V.C.110 for 1946–1947. MS.

————— 1961. Field meetings, 1959. Isles of Lewis and Harris. *Proc. B.S.B.I.* **4**: 204–206.

Campbell, R. N. 1974. St Kilda and its sheep. In *Island survivors: the ecology of the Soay Sheep of St Kilda* (Eds. **P. A. Jewell, C. Milner & J. M. Boyd**): 8–35. Athlone Press, London.

Cannon, J. F. M. 1978. History of plant recording on Mull. In *The island of Mull* (Eds. **A. C. Jermy** & **J. A. Crabbe**). British Museum (Natural History), London.

Carmichael, A. 1972. *Carmina Gadelica: Ortha nan Gaidheal*, 5 vols. Scottish Academic Press, Edinburgh & London (reprint of 2nd edition; First publ. 1900).

Chamberlain, D. F., King, R. A., McKean, D. R., Miller, A. G. & Nyberg, J. A. 1984. Freshwater macrophyte survey of selected lochs in the Uists. Royal Botanic Garden, Edinburgh. Internal report for N.C.C.

Cheke, A. S., & Reed, T. M. 1987. Flora of Berneray, Mingulay and Pabbay, Outer Hebrides in 1964. *Scot. Naturalist* **99**: 63–107.

**Clark, W. A.** 1938. The Flora of the Islands of Mingulay & Berneray. *Proc. Univ. Durh. Phil. Soc.* **10(1)**: 56–70.

———— 1939a. The Occurrence of *Arenaria norvegica* Gunn. and *Thlaspi alpestre* L. on Rhum (V.C.104) and *Carex Halleri* Gunn. on N. Uist. *Journ. Bot.* **77**: 4–5.

———— 1939b. Remarks on certain Hebridean Plants. *Vasculum*, **25**: 73–75.

———— 1939c. Noteworthy Plants from N. Uist, Baleshare, Monach Islands, Harris, Taransay, Mingulay & Berneray (V.C.110). *Proc. Univ. Durh. Phil. Soc.* **10**: 124–129.

———— 1942. A hybrid *Potamogeton* new to British lists. *Vasculum* **27**: 20.

———— 1943. Pondweeds from North Uist (V.C.110), with a special consideration of *Potamogeton rutilus* Wolfg. and a new hybrid. *Proc. Univ. Durh. Phil. Soc.* **10**: 368–373.

———— 1956. Plant distribution in the Western Isles. *Proc. Linn. Soc.* **167**: 96–103.

**Clark, W. A. & Harrison, J. W. H.** 1940. Noteworthy Plants from Great and Little Bernera (Lewis), Pabbay and Berneray (Harris) and the Uig district of Lewis. *Proc. Univ. Durh. Phil. Soc.* **10**: 214–221.

**Conacher, E. R. T.** 1980. Report. Field meeting, Barra, 8–15 July, 1978. *Watsonia* **13**: 88–89.

**Cooper, D.** 1970. *Skye*. Routledge & Kegan Paul, London.

———— 1979. *Road to the isles, travellers in the Hebrides, 1770–1914*. Routledge & Kegan Paul, London.

**Copping, A.** 1977. Field Excursion Report: Stornoway, Outer Hebrides 2–9 August, 1976. *Watsonia* **11**: 278–279.

———— 1978. *Phalaris arundinacea* in Barvas. *Watsonia* **12**: 158.

**Corrigan, D.** 1983. Granny Really Did Know Best, *Sunday Standard*, April 17, 1983, Glasgow.

**Coward, M. P.**, 1972. The Eastern Gneiss of South Uist. *Scott. J. Geol.*, **8**: 1 –12.

———— 1973. Heterogeneous deformation in the development of the Laxfordian complex of South Uist. *J. Geol. Soc. Lond.* **129**: 139–160.

**Coward, M. P. & Graham, R. H.**, 1973. The Laxfordian of the Outer Hebrides. In: *The Early Precambrian of Scotland and Related Rocks of Greenland* (Eds. **R. G. Park & J. Tarney**): 85–93. Keele.

**Crabbe, J. A., Jermy, A. C., & Walker, S.** 1970. The Distribution of *Dryopteris assimilis* S. Walker in Britain. *Watsonia* **8**: 3–13.

**Crummy J. A.** 1982. An expedition to Little Bernera. *Hebridean Naturalist* **6**: 38–43.

**Cunningham, P.** 1978. The Castle Grounds: The Stornoway Woods. The Stornoway Trust.

**Currie A.** 1977. Butterbur (*Petasites hybridus*) as an invading species on Machair in North-west Lewis. In *Sand Dune Machair* (Ed. **D. S. Ranwell**) **2**: 22–23. Institute of Terrestrial Ecology. (N.E.R.C).

———— 1979. The vegetation of the Outer Hebrides. *Proc. Roy. Soc. Edinb.* **77B**: 219–265.

———— 1981a. The vegetation of Gulleries on Moorland in Lewis, Outer Hebrides. *Hebridean Naturalist* **5**: 41–49.

———— 1981b. Vegetation on Islands of West Loch Roag, Lewis. *Hebridean Naturalist*, **5**: 57–58.

———— 1981c. The plants of the Outer Hebrides. *Scottish Wildlife* **17**: 13–18.

**Dahl, E.** 1954. Weathered gneisses at the island of Runde, Sunnmre, western Norway, and their geological interpretation. *Nytt Magasin for Botanikk* **3**: 5–23.

———— 1955. Biogeographic and geologic indications of unglaciated areas in Scandinavia during the glacial ages. *Bulletin of the Geological Society of America* **66**: 1499–1519.

———— 1959. Amfiatlantiske planter. *Blyttia* **16**: 93–121.

———— 1987. The nunatak theory reconsidered. *Ecological Bulletins* **38**: 77–94.

**Dandy, J. E. & Taylor, G.** 1940. Studies of British Potamogetons, XIV, *Potamogeton* in the Hebrides (vice-county 110). *J. Bot.* **78**: 139–147.

———— & ———— 1941. Studies of British Potamogetons. – XV. Further records of *Potamogeton* from the Hebrides. *J. Bot.* **79**: 97–101.

———— & ———— 1942. The identification of some Hebridean Potamogetons. *J. Bot.* **80**: 21–24.

———— & ———— 1944. Studies of British Potamogetons. – XVII. Further remarks on *Potamogeton berchtoldii*. *J. Bot.* **80**: 121–124.

**Darling, F.F.** 1939. North Rona: A North Atlantic Island. Royal Institution of Great Britain. (Friday, Jan: 20, 1939.) Paper from: *Proc. Roy. Inst.* Vol. x, Pt.iii.

———— 1947. Natural History of the Highlands & Islands. Collins (New Naturalist), London.

**Darling, F. F. & Boyd, J. M.** 1964. The Highlands and Islands. 2nd edition. Collins (New Naturalist), London.

**David, R. W.** 1982. The Distribution of *Carex maritima* Gunn. in Britain. *Watsonia* **14**: 178–180.

**Davidson, C. F.**, 1943. The Archaean rocks of the Rodel district, South Harris, Outer Hebrides. *Trans. Roy. Soc. Edinb.* **61**: 71–112.

**Davies, E. W.** 1952. Notes on *Carex flava* and its allies. 1. A sedge new to the British Isles. *Watsonia* **3**: 66–69.

**Dawson, J. L.** 1936. Natural History of Barra, Outer Hebrides. *Proc. Roy. Phil. Soc. Edinb.* **22**: 254–255.

**Dearnley, R.**, 1962. An outline of the Lewisian complex of the Outer Hebrides in relation to that of the Scottish mainland. *Q. J. Geol. Soc. Lond.* **118**: 143–166.

———— 1963. The Lewisian complex of South Harris, with some observations on the metamorphosed basic intrusions of the Outer Hebrides, Scotland. *Q. J. Geol. Soc. Lond.* **119**: 243–312.

**Dickinson, G.** 1968. The vegetation of Grogary machair, South Uist. Unpublished report for the Nature Conservancy (Scotland).

**Dickinson, G. & Randall, R. E.** 1979. An interpretation of machair vegetation. *Proc. Roy. Soc. Edinb.* **77B**: 267–278.

**Druce, G. C.** 1901. On three species of Rhinanthus new to the Scottish Flora. *Ann. Scott. Nat. Hist.* 177–178.

———— 1929. The report of the Secretary & Treasurer pp. 601–2. Plant Notes p. 612 etc. New Country & other records p. 722 etc. *B.E.C. 1928 Rep.*

———— 1932. *The comital Flora of the British Isles.* Buncle, Arbroath.

**Duncan, N., Bullock, D., & Taylor, K.** 1981. A report on the ecology and natural history of Boreray, St. Kilda. Dept. of Zoology, Univeristy of Durham.

**Duncan, W. S.** 1896. *Elatine hexandra* DC in the Outer Hebrides. *Ann. Scot. Nat. Hist for 1896*: 63–64.

**Dwelly, E.** 1973. *The Illustrated Gaelic-English Dictionary.* Gairm Publications, Glasgow (8th edition).

**Edees, E. S.** 1973. Notes on British Rubi, 1. *Watsonia* **9**: 247–251.

**Edees, E. S.** 1975. Notes on British Rubi, 3. *Watsonia* **10**: 331–343 (ref. on p.333).

**Edwards, K. J.** 1979. Earliest fossil evidence for *Koenigia islandica* – middle Devensian interstadial pollen from Lewis, Scotland. *Journal of Biogeography* **6**: 375–377.

**Elton, C. S.** 1938. Notes on the ecological & natural history of Pabbay [Harris]. *J. Ecol.* **26**: 275–97.

**Erdtman, G.** 1924. Studies in the micropalaeontology of postglacial deposits in northern Scotland and the Scotch Isles, with especial reference to the history of woodlands. *J. Linn. Soc.* **46**: 449–504.

**Ewing, P.** 1890–1895. A contribution to the topographical botany of the west of Scotland. *Trans Nat. Hist. Soc. Glasgow* **2**: 309–321 (1890), **3**: 159–165 (1890), **4**: 199–214 (1895).

———— 1892 & 1899. The Glasgow Catalogue of Native & Established Plants. 1st & 2nd edition.

**Fernald, M. L.** 1932. The linear-leaved North American species of *Potamogeton*, section *Axillares*. *Memoirs of the American Academy of Arts and Sciences* **17**: 1–183.

**Fisher, J.** 1956. Rockall. London, Bles.

**Fitzpatrick, E. A.** 1963. Deeply weathered rock in Scotland, its occurrence, age and contribution to soils. *J. Soil Sci.* **14**: 33–43.

**Flenley, J. R., Maloney, B. K., Ford, D., & Hallam, G.** 1975. *Trapa natans* in the British Flandrian. *Nature,* London **257**: 39–41.

**Fletcher, H. R.** 1959. Exploration of the Scottish flora. *Trans. Bot. Soc. Edinburgh* **38**: 30–47.

**Flinn, D.** 1978. The glaciation of the Outer Hebrides. *Geological Journal* **13**: 195–199.

**Forrest, J. E., A. R. Waterston & E. V. Watson** (Eds.) 1936. The Natural History of Barra. *Proc. Royal Physical Soc. of Edinb.* **22 (5)**: 241–296.

**Francis, P. W. & Sibson, R. H.**, 1973. The Outer Hebrides Thrust. In: *The Early Precambrian of Scotland and Related Rocks of Greenland* (Eds. **R. G. Park & J. Tarney**): 95–104. Keele.

**Geddes, A.** 1936. Lewis. In *Scot. Geogr. Mag.* L11, 217–231, 300–313. (with map of vegetation).

**Geikie, J.** 1894. *The great ice age and its relation to the antiquity of man.* 3rd edition. London, Stamford.

**Gibson, A. H.** 1891. Phanerogamic flora of St. Kilda. *Trans. Bot. Soc. Edinb.* **19**: 155.

**Gilbert, O. L., Holligan, P. H. & Holligan, M. S.** 1973. The flora of North Rona 1972. *Trans. Proc. Bot. Soc. Edinb.* **42**: 43–68.

**Gilbert, O. L. & Holligan, P. M.** 1979. *Puccinellia capillaris* (Liljebl.) Jans. × *P. maritima* (Huds.) Parl. on North Rona, Outer Hebrides. *Watsonia* **12**: 338–9.

**Gilbert, O. L. & Wathern, P.** 1976. The flora of the Flannan Isles. *Trans. Bot. Soc. Edinb.* **42**: 487–503.

**Gill, J. J. B.** 1971. *Cochlearia scotica* Druce – Does it exist in northern Scotland? *Watsonia* **8**: 395–6.

**Gill, J. J. B, McAllister, H. A, & Fearn, G. M.** 1978. Cytotaxonomic studies on the *Cochlearia officinalis* L. group from inland stations in Britain. *Watsonia* **12**: 15–21.

**Gimingham, C. H., Gemmell, A. R., & Grieg-Smith, P.** 1948. The vegetation of a sand-dune system in the Outer Hebrides. *Trans. Bot. Soc. Edinb.* **35**: 82–96.

**Gladstone, J.** 1928. Notes on the Flora, pp. 77–79 in *St Kilda* (Ed. **J. Mathieson**). *Scot. Geog. Mag.* **44 (2)**: 65–90.

**Glentworth, R.** 1979. Observations on the soils of the Outer Hebrides. *Proc. R. Soc. Edinb.* **77B**: 123–137.

**Gloyne, R. W.** 1968. Some climatic influences affecting hill land productivity. *Hill land productivity* (Ed. **I. V. Hunt**): 9–15. British Grassland Symposium No.4, Aberdeen.

**Goode, D. A. & Lindsay, R. A.** 1979. The peatland vegetation of Lewis. *Proc. Roy. Soc. Edinb.* **77B**, 279–293.

**Goodrich-Freer, A.** 1902. *Outer Isles.* Archibald Constable & Co. (Westminister) pp. xv and 448.

**Gordon, S.** 1923. *Hebridean Memories.* Cassell & Co. London pp. xii and 179, T63.

**Green, F. W. H.** 1964. The climate of Scotland. In *The vegetation of Scotland* (Ed. **J. H. Burnett**): pp. 15–35. Oliver & Boyd, Edinburgh.

**Green, F. W. H. & Harding, R. J.** 1983. Climate of the Inner Hebrides. *Proc. Roy. Soc. Edinb.* **83B**: 121–140.

**Greig-Smith, P., Gemmell, A. R., & Gimingham, C. H.** 1947. Tussock Formation in Ammophila arenaria (L.) Link. *New Phytol.* **46**: 262–268.

**Groves, E. W.** 1958. *Hippophae rhamnoides* in the British Isles. *Proc. B.S.B.I.* **3**. '110. Outer Hebrides – Without precise locality (Ewing, 1899)'.

**Gwynne, D., Milner, C. & Hornung, M.** 1974. The Vegetation & Soils of Hirta. In *Island survivors* (Eds. **P. A. Jewell, C. Milnes & J. Morton Boyd**). Athlone Press.

**Hall, P. M.** 1937. The Irish Marsh Orchids. *B.E.C. Report* **11**: 330–354.

**Hardy, M.** 1905. Esquisse de la geographie et de la Vegetation des Highlands d'Ecosse (with map of vegetation) [Paris].

**Harrison, H. H.** 1939. Botanical Investigations in the Hebrides. *Proc. King's College Agric. Students Assoc.*

**Harrison J. H.** 1948. Recent Researches on the Flora & Fauna of the Western Isles of Scotland and their Biogeographical significance. *Proc. Belfast N. H. & Phil. Soc.*, Ser. 2, **3**: 87–96.

———— 1949a. Field Studies in Orchis L. I. The structure of Dactylorchid populations on certain islands in the Inner and Outer Hebrides. *Trans. Bot. Soc. Edinb.* **35**: 26–66.

———— 1949b. A new orchid hybrid. *Vasculum* **34**:22

———— 1953. The North American and Lusitanian elements in the flora of the British Isles. In *The Changing Flora of Great Britain* (Ed. **J. E. Lousley**): 105–123. B.S.B.I., Arbroath: Buncle.

**Harrison, J. W. H.** 1938a. New Plants from the Outer Hebrides. *Vasculum* **24**: 116–117.

———— 1938b. Vasculum **24**: 116–117. (*Rosa sherardii* var. *cookei. Lonicera periclymenum* var. *clarki. Anacamptis pyramidalis* var. *fudayensis.*)

———— 1939a. The Hebridean form of the spotted Orchid, *Orchis fuchsii* Druce. *Vasculum* **25**: 109–112.

———— 1939b. Fauna & Flora of the Inner and Outer Hebrides, *Nature* **143**: 1004–1007.

———— 1940a. A day's collecting in the Isle of S. Uist. *The Entomologist* **73**: 1–4.

———— 1940b. More Hebridean Days. 11: Isles of Benbecula. *Entomologist* **74**: 1–5.

———— 1941a. A Preliminary Flora of the Outer Hebrides. *Proc. Univ. Durh. Phil. Soc.* **10**: 228–272.

———— 1941b. Flora & Fauna of the Inner & Outer Hebrides. *Nature* **147**: 134.

———— 1944. Hebridean Potamogetons once more. *Occasional notes from the Department of Botany, King's College, Newcastle upon Tyne*, series 2 **5**: 1–6.

———— 1945. Noteworthy sedges from the Inner and Outer Hebrides with an account of two species new to the British Isles. *Trans. Proc. Bot. Soc. Edinb.* **34**: 270–277.

———— 1948. Introduced vascular plants in the Scottish Western Isles. *N. W. Naturalist* **23**: 132–135.

———— 1948b. The passing of the ice age and its effect upon the plant and animal life of the Scottish Western Isles. In *The New Naturalist,* a Journal of British Natural History: 83–90. Collins, London.

———— 1949. Potamogetons in the Scottish Western Isles, with some remarks on the general natural history of the species. *Trans. & Proc. Bot. Soc. Edinb.* **35**: 1–25.

———— 1950a. A dozen years biogeographical research in the Inner and Outer Hebrides. *Proc. Univ. Durh. Phil. Soc.* **10**: 516–524.

———— 1950b. A pondweed, new to the European flora, from the Scottish Western Isles, with some remarks on the phytogeography of the island group. *Phyton. Annales Rei Botanicae* **2**: 104–109.

———— 1951a. Vascular Plants of the Outer Hebrides in 1950. *Proc. Univ. Durh. Phil. Soc.* **11**: 1–11.

———— 1951b. Observations on the Flora of the Isle of Lewis, Isle of Harris and the Shiant Isles in 1951. *Univ. Durh. Phil. Soc. Proc.* **11**: 83–90.

———— 1951c. Introduced Vascular plants in the Scottish Western Isles. *North Western Naturalist,* 1948: 132.

———— 1952. Occurrence of the American pondweed, *Potamogeton epihydrus* Raf., in the Hebrides. *Nature* **169**: 548–549.

———— 1953. Observations on the flora of the Isle of Lewis, Isle of Harris and the Shiant Isles in 1952. *Proc. Univ. Durh. Phil. Soc.* **11**: 83–90.

———— 1954a. Botanical Investigations in the Isles of Lewis, Harris, Taransay, Coppay & Shillay in 1953. *Proc. Univ. Durh. Phil. Soc.* **11**: 135–142.

———— 1954b. Observations on the Vascular Plants of the Outer Hebrides made in 1954. *Proc. Univ. Durh. Phil. Soc.* **12**: 29–34.

———— 1956. Botanising in the Outer Hebrides in 1955 and 1956. *Proc. Univ. Durh. Phil. Soc.* **12**: 141–149.

———— 1956b. On field studies of the distribution of the plants and animals of the Scottish Western Isles. *Proc. Linn. Soc.* **167**: 103.

———— 1957. Botanical investigations in the Isles of Lewis, Harris and Great Bernera in 1957. *Proc. Univ. Durh. Phil. Soc.* **13**: 54–62.

**Harrison, J. W. H. & Blackburn, K. B.** 1946. The occurrence of a nut of *Trapa natans* L. in the Outer Hebrides, with some account of the peat bogs adjoining the loch in which the discovery was made. *New Phytologist* **45**: 124–131.

**Harrison, J. W. H. & Bolton, E.** 1938. The Rosa Flora of the Inner & Outer Hebrides & of other Scottish Islands. *Trans. & Proc. Bot. Soc. Edinb.* **33**: 424–431.

**Harrison, J. W. H. & Clark, W. A.** 1941a. Potamogetons on the Isle of Benbecula. *Occasional notes from the Department of Botany, King's College, Newcastle upon Tyne,* series 2 **2**: 1–3.

———— & ———— 1941b. Pondweeds in the Outer Hebrides. *Occasional notes from the Department of Botany, King's College, Newcastle upon Tyne,* series 2 **3**: 1–3.

———— & ———— 1942a. Hebridean Potamogetons and their identifications. *Occasional notes from the Department of Botany, King's College, Newcastle upon Tyne,* series 2 **4**: 1–4.

———— & ———— 1942b. A note on *Potamogeton* × *suecicus* Richt. *Occasional notes from the Department of Botany, King's College, Newcastle upon Tyne,* series 2 **4**: 4.

**Harrison, J. W. H., Clark, W. A. Harrison, H. H. & Cooke, R. B.** 1938. A Review of Recent Articles on the Flora of the Outer Hebrides. *Vasculum* **25**: 52–57.

**Harrison, J. W. H. & Harrison, J. H.** 1950a. The vascular plants of Stuley Islands, the Isles of Grimsay and Ronay, with some remarks on the flora of Benbecula, South Uist and Barra. *Proc. Univ. Durh. Phil. Soc.* **10**: 499–515.

———— & ———— 1950b. A contribution to our knowledge of the flora of the Isles of Lewis, Harris, Killegray and Ensay. *Trans. & Proc. Bot. Soc. Edinb.* **35**: 132–156.

———— & ———— 1951. Further observations on the plants of the Outer & Inner Hebrides. *Trans. Bot. Soc. Edinb.* **35**: 415–476.

**Harrison, J. W. H., Harrison H. H., & Clark, W. A.** 1941a. Observations on the Flora of the Isle of Harris. (V.C. 110) *J. Bot.* **79**: 164–169.

————, ———— & ———— 1941b. Preliminary Flora of the Outer Hebrides. *Proc. Univ. Durh. Phil. Soc.* **10**: 228–273.

**Harrison, J. W. H., Harrison, H. H., Clark, W. A. & Cooke, R. B.** 1942a. Further Observations on the Vascular Plants of the Outer Hebrides (V.C. 110). *Proc. Univ. Durh. Phil. Soc.* **10**: 358–368.

————, ————, ———— & ———— 1942b. Vascular Plants from the Isle of Rhum (V.C. 104) and the Isle of South Uist (V.C. 110). *J. Botany* **80**: 113–116.

————, ————, ———— & ———— 1951. Further Observations on Vascular Plants of the Outer & Inner Hebrides. *Trans. Bot. Soc. Edinb.* **35**: 415–426.

**Harrison, J. W. H., Harrison, H. H., Cooke, R. B., & Clark, W. A.** 1939. Plants New to or rare in the Outer Hebrides I. From South Uist, Eriskay & Fuday. *J. Bot.* **77**: 1–4.

**Harrison, J. W. H., & Morton, J. K.** 1951. Botanical investigations in the Isles of Raasay, Rhum (V.C. 104) Lewis & Harris (V.C 110). *Proc. Univ. Durh. Phil. Soc.* **11**: 12–23.

**Harrold, P.** 1978. A glabrous variety of *Sagina subulata* (Sw.) Presl in Britain. *Trans. Bot. Soc. Edinb.* **43**:1–5.

**Heron, R.** 1794. *General view of the natural circumstances of those isles ... of ... Hebrides.* Edinburgh.

**Hobbs, A. M.** 1988. Conservation of leafy liverwort-rich *Calluna vulgaris* heath in Scotland. In: *Ecological change in the uplands* (Eds. **M. B. Usher & D. B. A. Thompson**): 339–343. Special publication no. 7 of the British Ecological Society, Blackwell, Oxford.

**Holden, A. V.** 1961. Concentration of chloride in fresh waters and rain water. *Nature* **192**: 961–962.

**Hudson, G., Towers, W., Bibby, J. S. & Henderson, D. J.** 1982. *The Outer Hebrides: soil and land capability for agriculture.* The Macaulay Institute for Soil Research, Aberdeen.

**Huiskes, A. H. L.** 1979. Biological Flora of the British Isles. *Ammophila arenaria. J. Ecol.* **67**: 363–382.

**Hulme, P. D.** 1985. The peatland vegetation of the Isle of Lewis and Harris and the Shetland Isles, Scotland. *Aquilo Ser. Bot.* **21**: 81–88.

**Hulme, P. D. & Blyth, A. W.** 1984. A classification of the peatland vegetation of the Isle of Lewis and Harris, Scotland. *Proc. 7th International Peat Congress*, Vol. 1. Dublin.

**Jehu, T. J. and Craig, R. M.**, 1923. Geology of the Outer Hebrides. Part I. The Barra Isles. *Trans. Roy. Soc. Edinb.* **53**: 419–441.

———— & ———— 1925. Geology of the Outer Hebrides. Part II. South Uist and Eriskay. *Trans. Roy. Soc. Edinb.* **53**: 615–641.

———— & ————.1926. Geology of the Outer Hebrides. Part III. North Uist and Benbecula. *Trans. Roy. Soc. Edinb.* **54**: 467–489.

———— & ———— 1927. Geology of the Outer Hebrides. Part IV. South Harris. *Trans. Roy. Soc. Edinb.* **55**: 457–488.

———— & ———— 1934. Geology of the Outer Hebrides. Part V. North Harris and Lewis. *Trans. Roy. Soc. Edinb.* **57**: 839–874.

**Jermy, A. C., Chater, A. O. & David R. W.** 1982. *Sedges of the British Isles.* B.S.B.I. Handbook No. 1.

**Jermy, A. C., Arnold, H. R., Farrell, L. & Perring, F. H.** (Eds) 1978. *Atlas of Ferns* of the British Isles. B.S.B.I. and B.P.S., London.

**Jewell, P. A., Milner, C. & Boyd, J. M.** 1974. *Island survivors: the ecology of the Soay sheep of St. Kilda.* Athlone Press, London.

**Jones, E. M.** 1975. Taxonomic studies of the genus Atriplex (Chenopodiaceae) in Britain. *Watsonia* **10**: 233–251 (ref. p.235).

**Keatinge, T. H. & Dickson, J. H.** 1979. Mid-Flandrian changes in vegetation on Mainland Orkney. *New Phytol.* **82**: 585–612.

**Kenneth, A. G., Lowe, M. R. & Tennant, D. J.** 1988. *Dactylorhiza lapponica* (Laest. ex Hartman) Soó in Scotland. *Watsonia* **17**: 37–41.

**King's College.** 1939. Remarks on the flora of the islands of the Barra Group. *Vasculum* **25**: 120–122.

**Landwehr, J.** 1977. *Wilde Orchideaen van Europe.* Ver. Behourd Natuurmovmenten Nederland. S'graveland.

**Lang, D.** 1980. *Orchids of Britain*. Oxford.

**Law, D. & Gilbert, D.** 1986. Saltmarsh survey N.W. Scotland – The Western Isles. N.C.C. internal report.

**Lowe, J. J. & Walker, M. J. C.** 1986. Late-glacial and early Flandrian environmental history of the Isle of Mull, Inner Hebrides, Scotland. *Trans. Roy. Soc. Edinb.: Earth Sciences* **77**: 1–20.

**Lowe, M. R.** 1985. References to *Dactylorhiza majalis, D. purpurella majaliformis* in Scotland. List of hybrids; etc. MS.

**MacCulloch, J.** 1819. A description of the western isles of Scotland, including the Isle of Man. Two vols. & atlas. London.

————— 1824. *The highlands and western isles of Scotland; . . . founded on a series of annual journeys between the years 1811 & 1821 . . .* Four vols. Longman, London.

**MacDonald, N.** 1953. The School of Scottish Studies Archives, tape accessioned SA 1953/21 B6, University of Edinburgh.

**Macgillivray, J.** 1842. An account of the island of St. Kilda, chiefly with reference to its natural history; from notes made during a visit in July 1840. *Edinb. Philosoph. Journ.* **32**, 47, 178.

**Macgillivray, W.** 1830. Account of the series of Islands usually denominated the Outer Hebrides. Edinb. *Journ. Nat. Geog. Sci.* **1**: 245–250, 401–411 & **2**: 87–95, 160–165, 321–334.

————— 1831. Report on the Outer Hebrides, *Trans. Highland Soc.*, New Series, **2**.

————— 1901. A memorial tribute to William MacGillivray. Printed privately, Edinburgh.

**Mackenzie, C.** 1936. *The Book of Barra*. Routledge.

**Mackintosh, J. & Urquhart, A.** 1983. Survey of the haymeadows of the Uists. Internal report, N.C.C.

**MacLeod, A. M.** 1949. Some aspects of the Plant Ecology of the Island of Barra. *Trans. Bot. Soc. Edinb.* **35**: 67–81.

**Malloch, A. J. C. & Okusanya, O. T.** 1979. An experimental investigation into the ecology of some maritime cliff species. I. Field observations. *J. Ecol.* **67**: 283–292.

**Manley, G.** 1948. The climate of the Hebrides. *New. Nat. J.* **1**: 77–82.

————— 1979. The climatic environment of the Outer Hebrides. *Proc. Roy. Soc. Edinb.* **77B**: 47–59.

**Martin, M.** 1698. *A late voyage to St. Kilda, the remotest of all the Hebrides*. London.

————— 1703. *A Description of the Western Isles of Scotland*. Mackay's reprint. 1934. Stirling.

**Mather, A. S. & Ritchie, W.** 1977. *The beaches of the Highlands and Islands of Scotland*. Edinburgh: Countryside Commission of Scotland.

**Mathieson, J.** 1928. St. Kilda (including 'Notes on the Flora', by J. Gladstone). *Scot. Geog. Mag.* **44**: 65.

**McVean, D. N.** 1961. Flora & vegetation of the Islands of St Kilda & North Rona in 1958. *J. Ecol.* **49**: 39–54.

**McVean, D. N. & Ratcliffe, D. A.** 1962. *Plant communities of the Scottish Highlands*. Monographs of the Nature Conservancy, 1. H.M.S.O., London.

**Meikle, R. D.** 1952. *Salix calodendron* Wimm. in Britain. *Watsonia* **2**: 243–248.

————— 1984. *Willows and Poplars of Great Britain and Ireland*, B.S.B.I. Handbook No.4.

**Meinertzhagen, R.** 1947 & 1953. Typed list of plants in N. & S. Uist. MSS.

**Milligan, S. F.** 1899. Scottish Archeological Cruise to the Inner & Outer Hebrides & far North. *Roy. Soc. Antiquaries of Ireland*, 1899.

**Milne-Redhead, E.** 1984. Maybud Sherwood Campbell (1903–1982), obituary. *Watsonia* **15**: 157–160.

**Monro, Sir D.** 1549. *A Description of the Western Isles of Scotland called Hybrides*. Eneas Mackay's reprint. 1934. Stirling.

**Murchison, R. I. & Geikie, A.**, 1861. On the altered rocks of the Western Islands of Scotland, and the North-western and Central Highlands. *Q. J. Geol. Soc. Lond.* **17**: 171–228.

**Murray, C.**. 1981. *Equisetum* × *trachyodon* in Skye, W. Scotland *Fern. Gaz.* **12 (3)**: 179–180.

**Myers, J. S.**, 1970. Gneiss types and their significance in the repeatedly deformed and metamorphosed Lewisian Complex of Western Harris, Outer Hebrides. *Scott. J. Geol.* **6**: 186–199.

————— 1971. The Late laxfordian granite-migmatite complex of western Harris, Outer Hebrides. *Scott. J. Geol.* **7**: 254–284.

**Naylor, R. E. L. & Cumming, R. H.** 1982. The recent spread of Butterbur in Lewis. *Trans. Bot. Soc. Edinb.* **44**: 31–34.

**N.C.C. Edinburgh** 1986. Agriculture and environment in the Outer Hebrides. 3 parts. Internal report.

**Nelmes, E.** 1947. Two critical groups of British sedges. *B.E.C. Report for 1945*: 95–105.

**Nelson, E.** 1976. *Monographie und Iconographie der Orchidaceen.* Gattung. III Dactylorhiza.

———— 1979. Nachtrage zu E. Nelson 1976. *Taxon* **28**: 592–593.

**Newton, A.** 1988. A new bramble from Skye and the Outer Hebrides. *Watsonia* 17: 173–174.

**Nicol, E. A. T.** 1936. The brackish-water lochs of North Uist. *Proc. Roy. Soc. Edinb.* 56: 169–195.

**Niven, W. M.** 1902. On the distribution of certain forest trees in Scotland, as shown by the investigation of post-glacial deposits. *Scot. Geog. Mag.* **18**: 24–29.

**Okusanya, O. T.** 1979. An experimental investigation into the ecology of some maritime cliff species. II. Germination studies. *J. Ecol.* **67**: 293–304.

**Padmore, P. A.** 1957. The varieties of *Ranunculus flammula* L. & the status of *R. scoticus* E. S. Marshall and of *R. reptans* L. *Watsonia* 4: 19–27.

**Page, C. N.** 1963. A hybrid horsetail from the Hebrides. *Brit. Fern Gaz.* **9 (4)**: 117–118. (Illustration of *Equisetum × dycei*).

———— 1981. A new name for a hybrid horsetail in Scotland. *Fern Gaz.* **9 (4)**: 178.

**Page, C. N. & Barker, M. A.** 1985. Ecology and geography of hybridisation in British and Irish horsetails. *Proc. Royal Soc. Edinb.* **86B**: 265–272.

**Paul, A. M.** 1987. The status of *Ophioglossum azoricum* (Ophioglossaceae: Pteridophyta) in the British Isles. *Fern. Gaz.* **13 (3)**: 173–187.

**Peacock, J. D.** 1984. Quaternary geology of the Outer Hebrides. British Geological Survey. Report 16 no.2. H.M.S.O. 26 pp.

**Pennant, T.** 1772 et seq. *A Tour in Scotland & Voyage to the Hebrides.* 1774–6 (with figures of plants). John Monk, Chester pp.viii and 379, T44.

**Perring, F. H.** 1953. A 17th Century contribution to the Scottish Flora. *Watsonia* 3: 36–40.

———— 1961. Report of field meeting to the Isle of Lewis & Harris, 1959. *Proc. Bot. Soc. Brit. Isles* 4, 204.

———— 1962. The Irish problem. *Proceedings of the Bournemouth Natural Science Society* 52: 1–13.

**Perring, F. H. & Farrell, L.** 1983. British Red Data Book, 1, vascular plants. Ed.2. Royal Soc. for Nature Conservation.

**Perring, F. H. & Randall, R. E.** 1972. An annotated flora of the Monach Isles National Nature Reserve, Outer Hebrides. *Trans. Bot. Soc. Edinb.* 41: 431–444.

**Perring, F. H. & Sell, P. D.** (Eds.) 1968. *Critical Supplement to the Atlas of the British Flora.* Nelson for B.S.B.I., London.

**Perring, F. H. & Walters, S. M.** (Eds.) 1976. *Atlas of the British Flora.* 2nd Ed. Nelson for B.S.B.I., London.

**Petch, C. P.** 1932. Additions to the flora of St. Kilda. *J. Bot.* **70**: 169–171.

———— 1933a. The vegetation of St. Kilda. *J. Ecol.* 21: 92–100.

———— 1933b. Petch-Willmott correspondence. MS.

**Philp, B.** 1983. Habitat survey of the Uists. Unpublished data, Scottish Wildlife Trust.

**Pitkin P., Barter G., Curry P., MacIntosh J., Orange A. & Urquhart U.** 1983. A botanical review of the S.S.S.I.'s of the north and west coasts of Uist and the nearby islands. N.C.C., Edinburgh.

**Poore, M. E. D. & Robertson, V. C.** 1949. The vegetation of St. Kilda in 1948. *J. Ecol.* **37**: 82–99.

**Poore, M. E. D. & McVean, D. N.** 1957. A new approach to Scottish mountain vegetation. *J. Ecol.* **45**: 401–439.

**Praeger, R. L.** 1897. Flora of St.Kilda (note) *Ann. Scot. Nat. Hist.* 1897: 53.

**Pritchard, N. M** 1950. Gentianella in Britain. II. *Gentianella septentrionale* (Druce) E. F. Warb. *Watsonia* 4: 218–237.

**Pugsley, H. W.** 1940. Notes on British Euphrasias. VI. *J. Bot.* **78**: 89–92.

———— 1942. New species of Hieracium in Britain. *J. Bot.* **79**: 177–183, 193–197.

———— 1948. *Prodromus of the British Hieracia.* Linnean Society, London (some V.C. records for spp).

**Randall, R. E.** 1973. Airborne salt deposition and its effects upon coastal plant distribution: the Monach Isles National Nature Reserve, Outer Hebrides. *Trans. Bot. Soc. Edinb.* 42: 153–162.

———— 1974. *Rorippa islandica* (Oeder) Borbas sensu strictu in the British Isles. *Watsonia* 10: 80–82.

———— 1976. Machair zonation of the Monach Isles National Nature Reserve, Outer Hebrides. *Trans. Bot. Soc. Edinb.* **42**: 441–462.

———— 1979. *Wild Flowers of the Hebrides.* Jarrold, Norwich.

**Ranwell, D. S.** (Ed.) 1974. *Sand Dune Machair* 1. Report of a seminar at Coastal Ecology Research Station, Norwich.

———— (Ed.) 1977. *Sand Dune Machair* 2. Institute of Terrestrial Ecology. (N.E.R.C.).

———— (Ed.) 1980. *Sand Dune Machair* 3. Report on meeting in the Outer Hebrides. Institute of Terrestrial Ecology. (N.E.R.C.).

**Ratcliffe, D. A.** 1968. An ecological account of Atlantic bryophytes in the British Isles. *New Phytol.* **67**: 365–439.

———— 1977a. *Highland Flora.* Highlands & Islands Development Board.

———— 1977b. *A nature conservation review.* Cambridge University Press.

**Rechinger, K. H.** 1961. Notes on *Rumex acetosa* L. in the British Isles. *Watsonia* **5**: 64–66.

**Rich, T. C. G. & Rich, M. D. B.** 1988. *Plant Crib.* B.S.B.I., London.

**Richardson, J.A.** 1970. John William Heslop Harrison (1881–1967), obituary. *Watsonia* **8**: 181–182.

**Ritchie, W.** 1966. The post-glacial rise in sea-level and coastal changes in the Uists. *Inst. of Brit. Geographers Trans. & Papers,* Publ. No. 39, 79–86.

———— 1979. Machair development and chronology in the Uists and adjacent islands. *Proc. Roy. Soc. Edinb.* **77B**: 107–122.

———— 1985. Intertidal and sub-tidal organic deposits and sea level changes in the Uists, Outer Hebrides. *Scott. J. Geol.* **21**: 161–176.

**Roberts, B., & Atkinson, R.** 1955. The Haskeir Rocks, North Uist. *Scot. Nat.* **69**: 9–18.

**Roberts, H. W. Kerr, D. H. & Seaton D.** 1959. The machair grasslands of the Hebrides. *J. Br. Grassl. Soc.* **14**: 223–228.

**Robertson, J. S.** 1984. *A Key to the Common Plant Communities of Scotland.* Macaulay Institute for Soil Research, Aberdeen.

**Rodwell, J.** (in press) *National Vegetation Classification.* University of Lancaster.

**Royal Botanic Garden, Edinburgh.** 1983. Survey of aquatic vegetation on South Uist and Benbecula. Internal report for N.C.C.

**Royal Botanic Garden, Edinburgh.** 1984. See Chamberlain et al.

**Royal Botanic Garden, Edinburgh.** 1985. See Biagi et al.

**Scott, T.,** 1894. Ferns from Barra (Outer Hebrides), *Ann. Scot. Nat. Hist.* 1894: 187 (*Polypodium phegopteris* in Barra above Sinclairs Lock).

———— 1895. The sea spleenwort in the island of Barra. *Ann. Scot. Nat. Hist.* 1895: 64.

**Scott W., Palmer R.** 1987. *The flowering plants and ferns of the Shetland Islands.* The Shetland Times, Lerwick.

**Sell, P. D.** 1967. Taxonomic and Nomenclatural notes on the British Flora. *Watsonia* **6**: 301–303.

**Shoolbred, W. A.** 1895. Plants Observed in the Outer Hebrides in 1894. *J. Bot.* **33**: 237–249.

———— in **Bennett, A.** 1896. Records for Scottish Plants additional to 'Topographical Botany' Ed 2 (1883). *Ann. Scot. Nat. Hist.* 1896 p. 115.

———— 1899. Notes on N. Uist. Plants, 1898. *J. Bot.* **37**: 478–481.

———— Records in *B.E.C. Reports,* 1894, 1895, 1898, 1899, 1900.

**Silverside, A. J.** 1977. A phytosociological survey of British arable-weed and related communities. Ph.D. thesis, University of Durham.

**Simpson, D.D.A.** 1976. The later Neolithic and Beaker settlement at Northton, Isle of Harris. *British Archaeological Report* **33**: 221–231.

**Slack, A. A.** 1986. Lightfoot and the exploration of the Scottish flora. Scottish Naturalist (B.S.B.I. conference report No. 20).

**Small, A.** (Ed.). 1979. *A St. Kilda handbook.* pp 95. University of Dundee, Dept. of Geography, Occasional Paper No. 5, and National Trust for Scotland.

**Smith, N. A.** 1987. Floral communities – Berneray [Barra]. *Royal Air Force Ornithological Society J.* **17**: 55–62.

**Smith, D. I. & Fettes, D. J.** 1979. The geological framework of the Outer Hebrides. *Proc. Roy. Soc. Edinb.* **77B**: 75–83.

**Smollett, T.** 1768. The present state of all nations.

**Snogerup, B.** 1982. *Odontites litoralis* subsp. *litoralis* in the British Isles. *Watsonia* **14**, 35–39.

**Soil Survey of Scotland** 1984. Organisation and methods: soil and land capability for agriculture 1:250 000 survey. The Macaulay Institute for Soil Research, Aberdeen.

**Somerville, A.** 1889. Notes on the Flora of the island of Barra. *Trans. Nat. Hist. Soc. Glasgow* **2**: 183–188.

———— 1890. Notes on the Flora of Barra & S. Uist. *Trans. Nat. Hist. Soc. Glasgow* **3**: 31–36.

———— 1891. Plants exhibited (1890) from Mingulay. *Nat. Hist. Soc. Glasgow Proc.* M.S. **3**: 47.

**Sparling, J. H.** 1967. The occurrence of *Schoenus nigricans* L. in blanket bogs. 1. Environmental conditions affecting the growth of *S. nigricans* in blanket bog. *J.Ecol.* **55**: 1–31.

**Spence, D. H. N.** 1960. Studies on the vegetation of Shetland. III. Scrub in Shetland and in South Uist, Outer Hebrides. *J. Ecol.* **48**: 73–95.

**Spence, D. H. N., Allen, E. D. & Fraser, J.** 1979. Macrophytic vegetation of fresh and brackish waters in and near the Loch Druidibeg National Nature Reserve, South Uist. *Proc. Roy. Soc. Edinb.* **77b**: 307–328.

**Stevens, A.** 1914. Notes on the geology of the Stornoway district of Lewis. *Trans. Geol. Soc. Glasg.* **15**: 51–63.

**Stirton, J.** 1885. On *Myurium hebridarum* and other Mosses in the Hebrides. *Scottish Naturalist* 1885: 182–183.

———— 1887. Account of a visit to the Outer Hebrides. *Proc. Nat. Hist. Soc. Glasgow* **1**: 55.

**Sutherland, D. G.** 1984. The Quaternary deposits and landforms of Scotland and the neighbouring shelves: a review. *Quaternary Science Reviews* **3**: 157–254.

**Sutherland, D. G., Ballantyne, D. K., & Walker, M. J. C.** 1984. Late Quaternary glaciation and environmental change on St Kilda, Scotland, and their palaeoclimatic significance. *Boreas* **13**: 261–272.

**Sutherland, D. G. & Walker, M. J. C. 1984.** A late Devensian ice-free area and possible interglacial site on the Isle of Lewis, Scotland. *Nature*, London **309**: 701–703.

**Tansley, A. G.** 1939. *The British Islands and their vegetation*, Cambridge University Press. pp. 930.

**Taschereau, P. M.** 1985. Taxonomy of *Atriplex* species indigenous to the British Isles. *Watsonia* **15**: 183–209.

**Thor, G.** 1988. The genus *Utricularia* in the Nordic countries, with special emphasis on *U.stygia* and *U.ochroleuca*. *Nordic J. of Botany* **3**: 213–225.

**Trail, J. W. H.** 1898–1909. Topographical Botany of Scotland, also additions & corrections. *Annals of Scottish Natural History*.

———— 1905. The Plants of the Flannan Islands. *Ann. Scot. Nat. Hist.* 1905: 187.

———— 1901. Euphrasia in Northern Scotland. *Ann. Scot. Nat. Hist.* 1901: 179.

**Turrill, W. B.** 1927. The Flora of St. Kilda. *B.E.C. Report* **8**: 428–444.

**von Weymarn, J.** 1974. Coastal development in Lewis and Harris, Outer Hebrides with special reference to the effects of glaciation. Unpublished Ph.D. thesis. University of Aberdeen.

**von Weymarn, J. A.** 1979. A new concept of glaciation in Lewis and Harris, the Outer Hebrides. *Proc. Roy. Soc. Edinb.* **77B**: 97–105.

**von Weymarn, J. & Edwards, K. J.** 1973. Interstadial site on the Island of Lewis, Scotland. *Nature*, London **246**: 473–474.

**Walford, T.** 1818. *The scientific tourist in England, Wales and Scotland.*

**Walker, J.** 1812. *An economical history of the Hebrides and highlands of Scotland.* 2 vols.

**Walker, M. J. C.** 1984. A pollen diagram from St Kilda, Outer Hebrides, Scotland. *New Phytol.* **97**: 99–113.

**Walker, M. J. C., Ballantyne, C. K., Lowe, J. J., & Sutherland, D. G.** 1988. A reinterpretation of the late-glacial environmental history of the Isle of Skye, Inner Hebrides, Scotland. *Journal of Quaternary Science* **3**: 135–146.

**Walters, S. M.** 1953. *Montia fontana* L. *Watsonia* **3**: 1–6.

———— 1988. A wild and garden *Sagina*. *B.S.B.I. News* **49**: 52.

**Waterston, A. R., Holden, A. V., Campbell, R. N. & Maitland, P. S.** 1979. The inland waters of the Outer Hebrides. *Proc. Royal Soc. Edinb.* **77b**: 329–351.

**Watson, E. V.** 1939a. Notes on the Flora of Barra, Outer Hebrides. *J. Bot.* **77**: 5–8.

———— 1939b. The Mosses of Barra, Outer Hebrides. *Trans. & Procs. Bot. Soc. Edinb.* **32**: 516–541 (with photograph of *Acer pseudoplatanus*).

**Watson, E. V. & Barlow, H. W. B.** 1936. The vegetation. In *The Natural History of Barra, Outer Hebrides*, by **Forrest et al**. *Proc. Roy. Phys. Soc.* **22**: 244–254.

**Watson, H. C.** 1847–1859. *Cybele Britannica* 4 vols. London.

———— 1873 & 1874. *Topographical Botany* 2 vols.

———— 1883. *Topographical Botany* 2nd edition. London.

**Watson, J.,** 1977. The Outer Hebrides: a geological perspective. *Proc. Geol. Ass.* **88**: 1–14.

**Watson, W. C. R.** 1949. Weihean species of Rubus in Britain. *Watsonia* **1**: 71–83.

**Whittington, G. & Ritchie, W.** 1988. *Flandrian environmental evolution on North-East Benbecula and southern Grimsay, Outer Hebrides, Scotland.* O'Dell Memorial Monograph, University of Aberdeen **21**, 46 pp.

**Wilkins, D. A. 1984.** The Flandrian woods of Lewis (Scotland). *J. Ecol.* **72**: 251–258.

**Wilmott, A. J.** 1938. *Carex spiculosa* var. *hebridensis* A. Benn. *J. Bot.* **76**: 137–141.

———— 1939. Notes on a short visit to Barra. *J. Bot.* **77**: 189–195.

———— 1941. A new variety of *Cerastium tetrandrum,* var. *pusillum* Wilmott. *J. Bot.* **79**: 102.

———— 1945. The vegetation of Uig. In *The Flora of Uig* (Ed. **M. S. Campbell**): 20–43. Buncle, Arbroath.

———— 1948. Further Botanising in Uig. *The Scot. Nat.* **60**: 82–90.

**Woodell, S. R. J. & Kootin–Sanwu, M.** 1971. Intraspecific variation in *Caltha palustris. New Phytol.* **70**: 173–186.

**Yeo, P. F.** 1978. A taxonomic revision of Euphrasia in Europe. *Bot. J. Linn. Soc.* **77**: 223–334.

# LIST OF SSSIs AND NNRs IN THE WESTERN ISLES as at 16.12.88

National Nature Reserves (these are also SSSI)

| NNR | Date Declared |
|-----|---------------|
| Loch Druidibeg | 13.3.58 |
| Monach Isles | 1.12.66 |
| North Rona & Sula Sgeir | 18.6.56 |
| St Kilda | 4.4.57 |

| SSSI | Area | Interest | Date | Grid Ref |
|------|------|----------|------|----------|
| Achmore Bogs, Lewis | 306.6 | B | 22.12.83 | NB314275 |
| Allt Volagir, S Uist | 22.5 | B | 27.1.84 | NF798292 |
| Baleshare & Kirkibost, N Uist | 1465.7 | B | 26.2.85 | NF785623 |
| Balranald Bog etc. N Uist | 838.1 | B | 6.12.84 | NF712705 |
| Bornish & Ormiclate Machair,SU | 662.7 | B | 16.11.88 | NF753309 |
| Flannan Isles, Lewis | 80.9 | B | 22.12.83 | NA692467 |
| Gress Saltings, Lewis | 91.3 | B | 12.9.84 | NB487414 |
| Howmore Estuary, etc. S Uist | 424.1 | J | 2.5.85 | NF756356 |
| Little Loch Roag Valley Bog | 19.2 | B | 22.12.83 | NB140250 |
| Loch a'Sgurr Pegmatite, Harris | 0.6 | G | 16.8.85 | NG070865 |
| Loch an Duin, N Uist | 3603.0 | J | 16.8.85 | NF935740 |
| Loch Bee, South Uist | 1172.9 | J | 30.10.84 | NF770430 |
| Loch Dalbeg, Lewis | 4.4 | B | 22.12.83 | NB227457 |
| Loch Druidibeg, South Uist | 1677.0 | J | 23.3.87 | NF782378 |
| Loch Hallan, South Uist | 338.4 | B | 4.2.88 | NF738224 |
| Loch Laxavat Ard, etc – Lewis | 255.5 | B | 22.12.83 | NB237386 |
| Loch Meurach, Harris | 2.2 | G | 16.8.85 | NG061877 |
| Loch na Cartach, Lewis | 22.6 | B | 23.1.84 | NB534499 |
| Loch nan Eilean Valley Bog | 31.8 | B | 29.1.84 | NB237233 |
| Loch Obisary, North Uist | 353.8 | B | .8.86 | NF896620 |
| Loch Orasay, Lewis | 96.6 | B | 22.12.83 | NB387283 |
| Loch Scadavay, North Uist | 526.5 | B | 18.7.84 | NF856686 |
| Loch Scarrasdale Valley Bog | 207.0 | B | 14.3.84 | NB493503 |
| Loch Stiapavat, Lewis | 35.5 | B | 14.3.84 | NB528643 |
| Loch Tuamister, Lewis | 8.2 | B | 22.12.83 | NB264456 |
| Lochs at Clachan, North Uist | 105.9 | B | 14.3.84 | NF810640 |
| Luskentyre Banks & Saltings | 1172.9 | J | 14.3.84 | NB080973 |
| Machairs Robach & Newton | 757.6 | J | 16.8.85 | NF873763 |
| Mangersta Sands, Lewis | 19.1 | G | 16.8.85 | NB009309 |
| Mingulay & Berneray | 819.0 | B | 22.12.83 | NL560830 |
| Monach Isles, North Uist | 577.0 | B | 12.12.83 | NF626623 |
| North Harris | 122920.6 | B | 12.9.84 | NB065115 |
| North Rona & Sula Sgeir | 130.0 | B | 22.12.83 | HW810324 |
| Northton Bay, Harris | 414.6 | J | 27.9.84 | NF990920 |
| Rockall, Harris | 0.7 | G | 27.1.84 | —— |
| Shiant Isles, Harris | 202.0 | J | 27.1.84 | NG418978 |
| Small Seal Islands: | 155.4 | B | 14.3.84 | |
| 1) Gasker, Harris | 27.7 | | | NA875115 |

| | | | | |
|---|---|---|---|---|
| 2) Coppay, Harris | 11.1 | | | NF933938 |
| 3) Haskeir, North Uist | 16.6 | | | NF615820 |
| 4) Shillay, Harris | 54.9 | | | NF880914 |
| 5) Causamul, North Uist | 12.0 | | | NF660705 |
| 6) Flodday, Barra | 33.2 | | | NL613923 |
| St Kilda, Harris | 835.0 | J | 14.3.84 | NA155050 |
| Stornoway Castle Woodlands | 208.7 | B | 27.1.84 | NB416330 |
| Tolsta Head, Lewis | 3.9 | G | 16.8.85 | NB557468 |
| Tong Saltings, Lewis | 417.2 | B | 14.3.84 | NB440358 |
| West Benbecula Lochs | 115.5 | B | 16.8.85 | NF771521 |

# Index

## Index to English names

## Index of Gaelic names

## Index to scientific names

CPSIA information can be obtained at www.ICGtesting.com
Printed in the USA
BVOW09s2259040615

402948BV00005B/29/P